T0294348

ANTHROPOCENE
CLIMATE CHANGE, CONTAGION, CONSOLATION

Poised, elegantly constructed poems which provide calm spaces for the reader to inhabit.
— CAROL ANN DUFFY, *former U. K. Poet Laureate*

I was very drawn to those photographs, especially the ones where books and papers took on a stratigraphic quality, sharp sedimentary layers of human words. — FORREST GANDER *on SS's 'Paper T[r]ails'*

A fine contemporary voice whose work is simultaneously urgent, intelligent, innovative, crafted and lyrical. — PETER PORTER

Sudeep Sen is a cosmopolitan poet par excellence, who knows how to interweave in his redefned and brilliant poetry, all the possible nuances of an imaginative global poetic language. —MAGDA CARNECI

In Fractals, *Sudeep Sen has created an impressively unified account of his long-term engagement with poetry as fusion of both passion and precision. Sen's international poetic reputation is founded on his distinct blend of personal, meditative lyricism with camera-sharp imagery and contemporary allusion. A vividly-focused, confident, comprehensive collection.* — JANE DRAYCOTT

Sudeep Sen is a truly international poet. In the era of globalisation, he has responded to the challenges of the connected world with a unique poetic synthesis. No other poet writing in English today manages to balance the steely North American tradition with the lyric sincerity to be found in much of the rest of the world, from the subcontinent to Europe and beyond. Sen responds uniquely to the artistic opportunities that have been opened up by the new global cultures. Sen's unusually creative response to our rapidly changing world makes him both innovative and exemplary. But this is not to forget that he is also simply a very fine, highly imagistic poet; one who produces such brilliant, tightly stitched pictures. The poetry is filmic rather than simply pictorial. This too marks Sen as a pioneer of the digital era. Sen is a highly discriminating poet. —FIONA SAMPSON in *The London Magazine*

Fractals gathers more than three decades of poetry from one of India's most technically gifted poets working in English. Sen's poetic scope and knowledge of form and metre will dazzle poetry lovers while drawing in more casual admirers with his sophistication of language and range of topics. Fractals is an important retrospective collection from a poet even now just reaching the height of his powers.
— *World Literature Today* (USA)

A poet of technical facility. —BRUCE KING in *The Oxford English Literary History*

Sen is an eclectic poet whose understated work eschews fashionable trends, while exhibiting considerable technical virtuosity and versatility. —JOHN THIEME in *Cambridge Guide to Literature in English*

Sen [has] extended the range of Indian verse in English to encompass a variety of alternative views of language, history and culture. —CHRIS COOK ed., *Pears Cyclopaedia* (Penguin)

A gifted poet .... I think everyone who works in Indian literature in English should thank him for all he has done. —DOM MORAES in *Sunday Midday*

In this artfully put together anthology [Aria], Sen translates from many languages. Sen's own translation of Jibanananda Das's 'Banalata Sen' is luminous enough to carry the entire book.
— ARSHIYA SATTAR in *Outlook* 'Best Books of the Year'

I read Rain with considerable admiration and pleasure. It is a word-perfect collection and its subject matter is both the measure of the rain and the spoken line.
—AMIT CHAUDHURI in *The Statesman* 'Best Book of the Year'

WORKS BY SUDEEP SEN

POETRY | FICTION | NON-FICTION
[Books | Chapbooks | CDs | E-books]
Leaning Against the Lamppost; The Man in the Hut; The Lunar Visitations; Kali in Ottava Rima; New York Times; Parallel: Selected Poems; South African Woodcut; Mount Vesuvius in Eight Frames; Dali's Twisted Hands; Postmarked India: New & Selected Poems; BodyText; Retracing American Contours; Almanac; Lines of Desire; A Blank Letter; Perpetual Diary; Postcards from Bangladesh; Monsoon; The Single Malt; Prayer Flag; Distracted Geographies: An Archipelago of Intent; Rain; Rainbow; Heat; Winter; Ladakh; Kargil; Fractals: New & Selected Poems | Translations 1980-2015; EroText; Anthropocene: Climate Change, Contagion, Consolation

DRAMA
[Radio | Stage]
Vesuvius *(London: BBC Radio)*, Vesuvius *(London: Border Crossing)*, BodyText *(London: Border Crossing)*, Rain *(New Delhi: British Council & Mumbai: Kala Ghoda Arts Festival)*, Wo|Man *(New Delhi: India International Centre)*, T⁵ *(New Delhi: The Attic)*, Rain *(New Delhi: India International Centre)*, EroText Replay

FILM
[Short Feature | Documentary]
Rhythm, White Shoe Story, Woman of a Thousand Fires, Babylon is Dying: Diary of Third Street, Flying Home, Prayer Flag, Lines of Desire, Silence, Love in the Time of Corona & others

TRANSLATION
[Books | Chapbooks]
In Another Tongue; Hayat Saif: Selected Poems; Love & Other Poems by Aminur Rahman; Spellbound & Other Poems by Fazal Shahabuddin; Love Poems by Shamsur Rahman; Aria | Anika; The Single Malt; Rain; Banyan; Mediterranean; Offering; DiaLogos; Indian Dessert; Mount Vesuvius; Incarnadine; Driftwood; Absences; Kaifi Azmi: Poems | Nazms; My Typewriter is My Piano: Selected Poems of Anamika, My Body is the Stepson of My Soul | Mera Sharir Meri Aatma Ka Sautela Beta: Selected Poems in Hindi Translation; Cowdust Hour | Godhuli Lagna: Selected Poems in Punjabi Translation

EDITOR
[Books | Journals | E-zine]
Lines Review Twelve Modern Young Indian Poets; Wasafiri Contemporary Writings from India, South Asia & the Diaspora; Index on Censorship Songs [Poems] of Partition; Biblio Portfolio of South Asian Poetry; The British Council Book of Emerging English Poets from Bangladesh; Dash: Four New German Writers; Sestet; Midnight's Grandchildren: Post-Independence English Poetry from India; The Literary Review Indian Poetry, The Yellow Nib Modern English Poetry by Indians, World Literature Today Writing from Modern India; Another English Poetry Foundation Anthology, Poetry Review Centrefold of Indian Poems; Prairie Schooner Feast Anthology of Indian Women's Poetry; Atlas; The HarperCollins Book of English Poetry; World English Poetry; Sahitya Akademi Modern English Poetry by Younger Indians, Contemporary English Poetry by Indians

ANTHROPOCENE
Climate Change, Contagion, Consolation

SUDEEP SEN

foreword by
CAROLYN FORCHÉ

PIPPA RANN
books & media
LONDON, UK

Published by
PIPPA RANN BOOKS & MEDIA
(an imprint of Salt Desert Media Group Limited)
7 Mulgrave Chambers, 26 Mulgrave Road
Sutton SM2 6LE, London, UK
publisher@pipparannbooks.com
www.pipprannbooks.com

Distributed by
PENGUIN RANDOM HOUSE INDIA in 17 Asian countries
GARDNERS BOOKS LIMITED in the UK & other countries
KINDLE Direct Publishing and others worldwide

Text & Cover Photo Copyright 2021 © SUDEEP SEN
Inside Photos Copyright 2021 © SUDEEP SEN
Other Photos Copyright 2021 © INDIVIDUAL PHOTOGRAPHERS
Author Photo Copyright 2021 © SIMRAN SINGH
www.sudeepsen.org
All rights reserved

ISBNs
978-1-913738-39-6 [hardback]
978-1-913738-36-5 [e-book]

NEW EDITION

No part of this book may be used or reproduced in any manner whatsoever
without the written permission of the publisher except in the case of
quotations embodied in critical articles or reviews.

Typeset in Cochin 11/13.2 pts. & Copperplate Gothic Bold 11/13.2pts

Printed and bound in Great Britain by Bell and Bain Ltd, Glasgow

ISBN 978-1-913738-39-6

for
SIMI
&
ANJU | CHAND

Ami grisha aakaashe'r dikhe dekhi —
dhulo dhaka megh gulo jeno
gorome kaapche — kothai jal o bristhi?
Megh gulo doorer mirage'r moton.
bristhi'r aasha — otai ek shotti.
— S.S., 'Bharateo Grisha'

I stare at summer's steaming sky —
layered thick in dust, clouds
quake in heat — bereft of moisture, rain.
Cloudbanks appear distant, mirage-like.
Yearning for rain — that is the only truth.
— S.S., 'Indian Summer'
Translated from the original Bengali by the poet

ANTHROPOCENE
/ˈanθrəpəˌsiːn/

adjective
relating to or denoting the current geological age, viewed as the period during which human activity has been the dominant influence on climate and the environment.
"we've become a major force of nature in this new Anthropocene epoch"

noun
the current geological age, viewed as the period during which human activity has been the dominant influence on climate and the environment.
"some geologists argue that the Anthropocene began with the Industrial Revolution"

CONTENTS

6

❖

CAROLYN FORCHÉ
'An Expansive, Urgent, Appeal to Humanity – A Beautiful Work of Lyric, Photographic Art'

1.

something still remains —
otherwise from ashes, smoke
would not rise again

('Ash Smoke')

With *Anthropocene: Climate Change, Contagion, Consolation*, Sudeep Sen has written a report from the frontline of an earth in peril, from India (and South Asia), countries that have already experienced what the rest of the world will endure in the decades to come if we do not shift course.

We read these poems as 'a 21st century epic', a form rarely attempted in our time — at once collective and personal, historical and immediate. I can think of few other contemporary poets who mastered this form, one of them being Sudeep's mentor, Derek Walcott, with his book-length epic, *Omeros*. It is fitting that Derek & Sigrid are also here in spirit: the poem, 'Driftwood', is dedicated to them.

In *Anthropocene*, we feel, viscerally, the heat in Delhi, "yolky," and oppressive, "like filigreed sand on the skin." In 'Afternoon Meltdown', we see that "Even stone slow-melts on the over-heated / tarmac. Smoke emanates from barks, // sharply crackling — incipient fire-flurries. / ... // Sky burns, heat spikes — life forms impaled. "We glimpse the skies, whether empty, beautiful or threatening, in poems and also in photographs taken from his terrace during the year's sequestration. We hear the palpable silence of the room where Sudeep composed this work in the deep solitude imposed by the pandemic.

2.

The images that Sudeep creates for this work are sharp-etched and tightly-wrought. I called it an 'epic', but it is really a collage of haiku, prose poems, meditations, montage lyric narratives, and photography, constituting a hybrid form that allows the poet to circle and imagine the forest of the situation in which we find ourselves. By including the front pages of *The New York Times*, with the names of the dead blacked-out, he ruptures the elegiac mode with a stark reminder of the ineffable enormity of pandemic grief. The elegiac tone extends to a world we are losing — other species, even kinds of weather — the shifts in the air and atmosphere, even the heat of summer, always experienced in Delhi, but now more intense and acute.

Sudeep's images become emblematic, a kind of portrait of the time that we can feel in our bodies, vividly sensorial. We feel the oppression of the heat and the relief of the breeze. He goes out to the terrace and takes photographs, and then he comes inside and worries about what is to come, with a deep awareness throughout of the collective experience of this moment.

It is for the first time in contemporary history that we are all afflicted at once with the same precise difficulty — which is to survive a novel virus to which we are, as a species vulnerable. I think this is the first pandemic and climate change book I have seen which addresses this.

Death knells peal, numbers multiply,
virus ravages us, one by one.

Newspaper columns loom, unsteady
ghostly apparitions on broadsheets —

name, age, date of death —
tall epitaphs in fine print.

(from 'Obituary')

I want to speak specifically of the poem, 'Obituary', which is part of the 'Pandemic' section of the book. It is, in a sense, the way we are mirrored back the deaths of the pandemic, in bureaucratic and reportorial language, journalistic language, those columns of newsprint, those names listed that become so dense that the screen becomes black.

The 1996 Nobel Laureate Polish poet, Wisalawa Symborska, once said that "history counts its skeletons in round numbers" — I think this is so as not to account for the pathos and complexity and difficulty and pain of individual death. It is essential to acknowledge individual lives —and I think that is what the poem is addressing.

It is the poets' work to challenge other language games, other narratives: the journalistic, the bureaucratic, even the scientific — and to deepen them and to focus our attention more spiritually and passionately on the experience of human subjectivity as I think Sudeep does in this book.

4.

Anthropocene is an expansive, urgent, appeal to humanity, and it is also a beautiful work of lyric and photographic art that concludes on a note of consolation, prayer and hope. Appropriately, the volume is book-ended with a quote from another Polish poet, Czeslaw Milosz:

> *My generation was lost. Cities too. And nations.*
> *But all this a little later.*
> *Meanwhile, in the window, a swallow.*

Sudeep Sen has confronted, with essential and moving lyric art, the state of the Earth in the 21st century. *Anthropocene* is a poet's report from a collapsing biosphere, summoning human courage to alter course.

— C. F.

May 21, 2021 | Washington DC

[NOTE: The Foreword comes from the remarks made by Carolyn Forché, during the international launch of the book.]

1 *Prologue*

MEDITATION

This grand show is eternal. It is always sunrise somewhere; the dew is never dried all at once; a shower is forever falling; vapour is ever rising. Eternal sunrise, eternal dawn and gloaming, on sea and continents and islands, each in its turn, as the round earth rolls.
— JOHN MUIR

I am prisoner of a gaudy and unlivable present, where all forms of human society have reached an extreme of their cycle and there is no imagining what new forms they may assume.
— ITALO CALVINO, *Invisible Cities*

SUDEEP SEN
'THE ROLE OF THE ARTIST IS NOT TO LOOK AWAY'

1.

I paint flowers so they will not die.
— FRIDA KAHLO

I have always been deeply fascinated with the natural world and its phenomenon. I have vivid memories of devouring issues of the *National Geographic* magazine as a child, collecting their wonderfully detailed fold-out maps, watching Carl Sagan and David Attenborough's documentaries on the cosmos, wildlife and our planet. I marvelled at how solar and lunar gravitational pulls choreograph the seasonal orchestra — how the wind's slipstreams dance, the oceans churn, and earth's tilted rotation creates the vicissitudes of tides.

There was a point when geography was my favourite subject at school. For a while, I even wanted to become an oceanographer — so I could explore the earth's deepest point in the Mariana Trench, and ride icebergs in the Arctic/Antarctic photographing the colour-blazed magnetic lights of Aurora Borealis/ Australis.

As an avid reader, I have been tuned in to the global discussion on climate change — this man-made tragedy that threatens our planet in what has become the critical age of anthropocene. Still, I could not help feeling shaken to the core, when I chanced upon a news clip featuring the President of the island nation of Kiribati (in the central Pacific Ocean) informing the rest of the world that the first country to be submerged would be theirs — and that their people would be the first 'climate refugees,' in the not-so-distant future.

Stanford University climatologist, Stephen H. Schneider, has said that the small island nation of Kiribati is made up of 33 small atolls, none of which is more than 6.5 feet above the South Pacific, and it is only a matter of time before the rising sea submerges the entire country. "For Kiribati, the tipping point has already occurred," Schneider told *The Washington Post* in January 29, 2006. "As far as they're concerned, it's tipped, but they have no economic clout in the world."

2.

Climate change poses a powerful challenge to what is perhaps the single most important political conception of the modern era: the idea of freedom, which is central not only to contemporary politics but also to the humanities, the arts and literature.
— AMITAV GHOSH, *The Great Derangement: Climate Change and the Unthinkable*

It is ironic that the most powerful nations of our world are the largest emitters of greenhouse gases and contribute the maximum tonnage of carbon footprint. The Covid-19 pandemic is only the most recent in a series of epidemics that have ravaged the world (both human and animal) in recent decades. With the melting of polar ice caps and the rapid thaw of glacial sites — humanity faces, among other dangers — discharge of dangerous levels of methane

& CO_2 gases, and the activation of infectious life forms that were frozen for millennia. Add to this, rising sea levels, increased frequency of natural disasters, a growing tribe of self-serving Fascist leaders — and life as we know it seems to be imploding.

Katie Pavid writes on the Natural History Museum website, "The Earth is 4.5 billion years old, and modern humans have been around for a mere 200,000 years. Yet in that time we have fundamentally altered the physical, chemical and biological systems of the planet on which we and all other organisms depend. In the past 60 years in particular, these human impacts have unfolded at an unprecedented rate and scale. This period is sometimes known as the 'Great Acceleration'. Carbon dioxide emissions, global warming, ocean acidification, habitat destruction, extinction and wide scale natural resource extraction are all signs that we have significantly modified our planet." Still, it seems clear the world is not taking sufficient heed.

3.

At what distance should I keep myself from others in order to build with them a sociability without alienation and a solitude without exile?
— ROLAND BARTHES, *How to Live Together*

I spend most of my waking hours in the day (and night) in my book-lined study. The panoramic picture window across my desk is the lens through which I view the changing of seasons imprinted on the magnificent wide-topped *neem* tree. The bough's intricate armature, the leaves' serrated floret-pattern, the tree's broccoli-shaped structure — all provide an exo-skeleton for my canvas — and the constantly-altering skyscape, provides a slideshow cyclorama.

I have been writing on climate related phenomenon for a long time. My poems frequently dwell on the theme of excess. Having lived in Delhi for most of my life and braved its predominantly hot weather for decades, I have often written on aspects of 'heat'. Heat annoys, repels, inspires and exasperates:

> *Heat outside is like filigreed sand on my skin —*
> *swift, sharp, pointed, deceptive, furnace hot."*
> (from 'Heat Sand')

In the early 2000s however, I lived in Bangladesh for some years. As a result of living for half a decade in the region of the 'two Bengals' — West Bengal in India and Bangladesh — I published a book titled *Monsoon* that was later republished as *Rain*.

> *It is bone-dry — I pray for any moisture that might fall from the emaciated skies — // There is a cloud, just a solitary cloud wafting perilously — // But it is too far in the distance for any real hope — for rain.*
> ('Drought, Cloud', *Rain*)

The book reflected and meditated on the various moods and effects of rain — its passion and politics, its beauty and fury, its hope and hopelessness, its ability of "douse and arouse". In some ways, it was a book on aspects of climate change — even though I confess that I did not overtly set out to do so, at least not consciously.

4.

> *[The pandemic] has reinforced my understanding of the best parts of human nature, and the worst. The best is the selfless courage of so many on the front lines, and the brilliance of scientists trying to find the cure. The worst is the degeneration of parts of society into an aggressive, hostile, ignorant, bigoted rabble. I've always believed that both these elements exist.*

The pandemic shines a bright light both on nobility and on ugliness, the will to overcome adversity and a sort of Lord of the Flies *barbarism.*
— SALMAN RUSHDIE, *Interview magazine* (June 1, 2020)

During the early days of the Covid-19 lockdown, things were changing so fast around us that it was viscerally affecting our society — the play of politics, the way people thought and reacted, the changing culture of 'working from home' for the privileged and lack of work for the dispossessed, the gruesome images of migrants walking hundreds of kilometres in the unforgiving weather riddled by hunger and pain, the quarantine, the virus — how can all these not affect you psychologically as well. To make matters worse, the pandemic was accompanied by floods, locust attack, earthquakes, and more.

So, when Raj Kamal Jha, the novelist, and editor of one of India's leading newspapers, *The Indian Express*, asked me to contribute something on the pandemic for the editorial pages, I expect he was looking for a sensitive and incisive non-fiction piece from an artist's point of view. I am pretty certain he did not expect my response in the form of poetry. In any case, I sent him the poem 'Love in the Time of Corona' — and I must say I was impressed that he chose to carry it in their weekend edition — "the first time that they had carried poetry" according to him. Thereafter, the poem took on a life of its own.

The former UK poet laureate Carol Ann Duffy selected it for a world project 'Write Where We Are Now,' currently hosted on the Manchester Metropolitan University's website. Among others, *Singing in Bad Times: A Global Anthology of Poetry Under Lockdown* (Penguin Random House), *The Calvert Journal* (UK), *ArtVirus* (USA) and many others, carried the poem.

In addition, a slew of translations appeared — in French, Spanish, Serbian, Persian, Urdu, Marathi, Bengali, and Hindi.

5.

Art does not reproduce the visible; rather, it makes visible.
— PAUL KLEE

As a poet and literary writer, and as a person who works from home, 'self-isolation' is nothing new, abnormal or unusual. Over the last three decades, I have spent most of my working hours happily and voluntarily self-isolated and quarantined, cocooned in the world of ideas, surrounded by books and literary artifacts in my office-study.

In the early days of the pandemic, with the country under a 'lockdown' and no transportation allowed on the streets, with offices and industries shut— Delhi started showing signs of regeneration. It was extraordinary how quickly we saw signs of nature healing itself — clean air, blue skies during the day, starlit skies at night, the skyscapes during dawn and dusk everyday utterly spectacular, and the constant elation of silence. Outside my window, the sparrows, the bees and the butterflies were back — and for a while it felt like the Delhi of yesteryears, the one I had grown up in the 1970s & 80s.

In this period, I would spend two to three hours in the evening on my terrace — reading, walking, watching, calling out to neighbours, eavesdropping on birds, "listening to the stars", and photographing skyscapes. I took to capturing the skies from exactly the same vantage point on my terrace day after day — and the selection of photographs in this book will give you an idea of the varied vivacity of the ever-changing canvas.

6.

The role of the artist is not to look away.
— AKIRA KUROSAWA

Nature has always inspired writers and artists. I have not been immune to this attraction either. However, I have noticed that with the passing of time, the celebration of nature in my poetry and prose has been tempered with warnings of what this irreversible change in climate means for the earth.

Amid all the clamour of public rhetoric and widespread distress, this book is a quiet artistic offering. It is a testament to our fervent times where a fascist political din overrides the silence of introspection, where the ravages of climate change scar humanity, where the cleaving schism between the rich and poor becomes ever-widening, where racism peaks at an all-time high, where toxicity amongst people proliferates, and fake news abounds.

In this book, you will experience the wider (and my personal) struggle with pollution and co-morbidities; the sharp rise and fall in atmospheric pressures, unusual heat spikes; unseasonal rain and hailstorm; invading oceans swallowing up coastlines around the world; floods; cyclones, devastation; illnesses — physical and psychological.

Even in the most spectacular sights, one sees the "terrible beauty" that is contained within. We know that the sunsets are more redolent due to pollution in the air — and that certain geological features are stunning because of the impurities they contain. In 'Akrotiri' on the volcanic Grecian island of Santorini, we see:

SUDEEP SEN

sand-soil compacted mineral / paintings — rainbow reserved normally for the skies.

Some pieces in the book refer only obliquely to these subjects, but I have included them since the references there are vital and important — they serve to underline and remind us of the constant, insidious, corrosion that is the accompanying chorus to life as we know it.

Holocene's carbon-footprint — its death text, unceasing.
(from 'Burning Ghats, Varanasi')

Several literary techniques and forms have been used to show our world's passage from utopia to dystopia — so you will see on display formal and free verse forms, prose poems, fragmented prose, flash and micro-fiction, and more.

"Everywhere I go, I find a poet has been there before me," Sigmund Freud had once remarked. *Anthropocene: Climate Change, Contagion, Consolation* while engaging with the most urgent topics that face humanity now — climate change and pandemic — is ultimately a prayer for positivity and hope. It is time again to slow down, to consume less, to love more selflessly and expansively.

Hope, heed, heal — our song, in present tense.
(from 'Love in the Time of Corona')

— S. S.
New Delhi, India | June 21, 2020 | Summer Solstice

❖

2

Anthropocene

CLIMATE CHANGE

Transformed utterly:
A terrible beauty is born.
—W B YEATS, 'Easter, 1916'

These little strokes whose syllables confirm
an altering reality for vision //
on the blank page, or the imagined frame
of a crisp canvas, are not just his own.
— DEREK WALCOTT, *Tiepolo's Hound*

I.E. [THAT IS]

for Aria

i.e.

   because you hear —
the sound
of a lone rustling leaf —
you hear the sea.

i.e.

   because I consider
the sea silent —
you hear its silence
in my studio.

i.e

   & because of that —
the silence will not empty
the sea
of its leaves.

DISEMBODIED
for Amitav Ghosh

1.

My body carved from abandoned bricks of a ruined temple,
                        from minaret-shards of an old mosque,
        from slate-remnants of a medieval church apse,
            from soil tilled by my ancestors.

My bones don't fit together correctly   as they should —
the searing ultra-violet light from Aurora Borealis
                patches and etch-corrects my orientation —
magnetic pulses prove potent.

My flesh sculpted from fruits of the tropics,
                                blood from coconut water,
skin coloured by brown bark of Indian teak.

My lungs fuelled by Delhi's insidious toxic air
                echo asthmatic sounds, a new vinyl dub-remix.
Our universe — where radiation germinates from human follies,
            where contamination persists from mistrust,
        where pleasures of sex are merely a sport —
where everything is ambition,
everything is desire,   everything is nothing.
                Nothing and everything.

2.

White light everywhere,
        but no one can recognize its hue,
no one knows that there is colour in it —   all possible colours.

Body worshipped, not for its blessing,
                but its contour —
                        artificial shape shaped by Nautilus.

Skin moistened by L'Oreal
              and not by season's first rains —
skeleton's strength not shaped by earthquakes
              or slow-moulded by fearless forest-fires.

Ice-caps are rapidly melting — too fast to arrest glacial slide.
        In the near future — there will be no water left
or too much water that is undrinkable,
                        excess water that will drown us all.
Disembodied floats,   afloat like Noah's Ark —

no GPS, no pole-star navigation,   no fossil fuel to burn away —
  just maps with empty grids and names of places that might exist.

Already, there is too much traffic on the road —
        unpeopled hollow metal-shells   without brakes,
swerve about   directionless — looking for an elusive compass.

❧

GLOBAL WARMING

Stillness — over-heated air sucks out everything —   still

              not   strong enough to create vacuum
                      to attract rain-clouds
in this low pressure   of high heat.

            Leeward-windward —   face-off in vain.
Rain where there never was,
                                no rain where there was.

Climate patterns   total disarray —   defiantly altered
                  weather systems   topsy-turvy —
global warming's   man-made havoc.

            Earthquakes — overground, underground,
undersea —
              destruction, death,   cyclone, flood,
                                pestilence, pollution.

Stillness, ever stiller still —   all still-born.

ʒ

RISING SEA LEVELS

The ever-hooded, tragic-gestured sea.
— WALLACE STEVENS

Granite outcrop
    that once jutted out
of the ebullient sea —

fifty metres
    from the shore —
is seen no more.

The lighthouse —
    our beacon,
an adventure island.

We'd swim to its base —
    shells, mussels,
striated oval stones,

seagull eggs nested
    on slippery
kelp-laced rocks.

As children, reality —
    as adults,
a submerged memory.

❧

CLIMATE CHANGE

1. YESTERDAY

Winter is fading —
  Delhi's February sun
slow-warming the air.

Time to peel off
  the extra woollen layer.
Time to shell again

freshly roasted peanuts —
  fingers cracking open
their uneven bark-skin.

After the season's severe
  cold, we prepare
for spring and summer,

airing out duvets
  before storing them
until next year —

rolled up in cupboard's
  mothball company.
Yesterday — was brief.

2. TODAY

Today, there is
    unexpected rain here —
unseasonal snow

amid chill-sharp wind
    in the northern hills.
The winter clothes

I had washed and hung
    out to dry, are now
wet and damp again.

Dogs are coiled
    around each other,
birds' feathers

puffed up for warmth,
    squirrels gone
back into hiding —

as we stare starkly
    at the climate change
we've helped create.

🦚

CLIMATE CHANGE 2

haiku

climate change: changes
the terrible beauty of
unbearable heat.

DROUGHT, CLOUD

It is bone-dry — I pray for any moisture
that might fall from the emaciated skies —

There is a cloud, just a solitary cloud
wafting perilously —

But it is too far in the distance for any real
hope — for rain.

POLLUTION

Neem's serrated leaves
   outside my study

wear season's toxicity
   on their exposed skin —

wan arteries choked,
   marking scant time.

Ash pollutants from
   city power plants,

crop-stubble smoke from
   burning fields —

pockmark, damage
   every visible vein —

as tree leaf-ridges
   struggle to supply life

to her pale-green lungs.
   Air is thick, heavy,

unclean, unworthy.
   Neem, once acted as

a filter for us,
   now needs one herself.

❧

ASPHYXIA

Even leaves on the trees of this *Unreal City*
fold, curl, bleed and weep, choking—

the air is dense, murky like stale lentil-soup —
like *the yellow fog that rubs its back*

upon the window-panes, / the yellow
smoke that rubs its muzzle muzzling

our breath, our words wheezing
in pulmonary distress. How many masks

do you need to mask the bloated AQI scale?
Callousness breed calluses in our lungs,

pollutants lacerate our larynx, breathlessness
wheeze. This deathly gas chamber,

this Capital melancholy, this burning crop,
this building dust, factory fumes, pestilence —

Sweet Yamuna *run softly till I end my*
smog — this insidious dirge, this unripe rot.

🦢

[NOTE: Lines/phrases in italics are from T S Eliot's poems:
'The Waste Land' and 'The Love Song of J Alfred Prufrock']

SUMMER HEAT

It's early in the season — summer's mere beginning. Yet, it is already searing hot. Heat melts everything — burns wood, tissue, metal and the human heart. Stones on the ground steam, simmer, ignite — tarmac slow-melts into a viscous black sea.

Neem tree branches in front of my study shrink like emaciated skeletal figures — an apocalyptic architectural shape — a haze, blanched wood-green, etiolated.

The power lines spark, short-circuit — no power for hours on end. Everyday it is the same. Speech on the telephone lines stutter — a new language of inconsequence. Even the clock hands find it hard to move — keeping time is unnatural in these times.

Tap water scalds everything it falls on — turning all furnace hot. Heat rises from everywhere — surfaces, terraces, walls, linen, food, water — everything is vaporous.

A mirage shimmering, a hallucinatory vacuum, a red-hot deception. Eyes are mere sockets, human skin rough leather, tongue a shriveled dry prune.

No moisture, just steam, heat, heat, and more heat. Barren. Everything seems in short supply, except heat.

drips ochre at 48°C,
drenched yolky heat.

Hotter it is, more
incandescent its colour —

sparking laburnums
to ignite, incinerate.

Heat — saturating
shrouds brighter —

dessicating our
throats, parched —

lungs heaving,
breathless, killing us

dry — burning,
yellow amaltas pyre.

Heat Sand

Heat outside is like filigreed sand on my skin —
swift, sharp, pointed, deceptive, furnace hot.

The dry atmosphere simmers, sears, scorches
living tissues — greens to dead brown —

melting street tarmac into viscous volcanic glue.
UV-rays refract swiftly through the dense air —

this city — a glistening glassy mirage, fuming,
fulminating, frothing — weather's deathly touch.

AFTERNOON MELTDOWN

Ah earth you old extinguisher.
— SAMUEL BECKETT, *Happy Days*

On days when it was too hot, ...
a heavy breeze brought the smell of tar.
— GUSTAVE FLAUBERT, *A Simple Heart*

Even stone slow-melts on the over-heated
tarmac. Smoke emanates from barks,

sharply crackling — incipient fire-flurries.
Looking for cooler earth under tree shade,

dogs dig deeper and deeper
into the dry, thinly available soil. Electrical

lines overhead glow, rutilant in heat.
Sheltering birds ruffle in uneasy agitation.

Temperatures soar, peaking new highs —
recorded meteorological indices shatter.

Our modest umbrellas fray,
flounder under the sun's ruthless exposure.

Sky burns, heat spikes — lifeforms impaled.

꒐

THE THIRD POLE

1.

Ice-caps constantly corroding,
   conspicuously melting the poles'

circumference, reducing their girth.
   I sit on the Dharamkot slopes —

watching the fading folded hills,
   cumulonimbus clouds veiling

Himalayas' towering snow peaks.
   Another polar crest, 'the third pole' —

like the older North-South —
   slow dissolves, thaws, deliquesces.

2.

The 14th Dalai Lama celebrates his 85th —
   another birthday in Indian exile.

His signature infectious cackle,
   a childlike laughter, a spontaneous

outburst — an oblique innocence.
   Dharamshala is a few hours away

on foot, through pine wood paths.
   Prayer chants waft. In this thin air

floats an immutable magic — a hope,
   perhaps, to arrest the glacial slide.

CONCRETE GRAVES

Arrogance, avarice,
real estate seduction —

sly filial deceit,
blighted brick buildings

spurring breath-wheeze,
asthma's bronchial blocks —

death-dust part-protected
by masks, Ventolin puffs —

more canned sprays,
less fresh air to breathe.

Skeletal skyscrapers,
unfinished flyovers

collapse prematurely
burying people —

none held responsible.
Darkly efficient, untimely —

a fast-track to
our planet's detonation.

ॐ

1

Grey drizzle's incessant mania —
wet competing with wet —

wiping out white light —
mirrors Himalayan snowmelt.

2

Slate-grey moist sheen —
a deceptive wet-suit façade

fluctuates and alters
with weather's polygamy.

3

Water-kettle simmers endlessly —
wet curls of black tea-leaves,

pungency of ginger, tulsi & honey
soothing throat's coarse timbre.

4

Nip in the air, infectious,
sneeze and wheeze

audible through the diaphragm's
buried emotion and flutter.

5

A wall-lizard sheds its tentative tail
moulting magic —

pigeons fluff, expand inwardly
into pregnant grey — coital hope.

6

Tree-leaves gain temporary shine
before dust settles —

smoke trails cut short desires —
heavy weather's vapour, thins.

7

Time will gather all rainwater,
harness its overflow, freeze

prism-droplets, so a new rainbow
refracts. But it's too wet — still.

꒳

RAIN CHARM

Another rain swept day leaves everything water-logged — ponds, drains, streets, and rivers — everywhere water is overflowing. The green blades of grass in the garden lie submerged under a rippling shallow sheet of water. Through refraction, they take on magical underwater seaweed shapes. Except here, the grass is evenly cropped, so it appears as a glazed woven mat of wet-green. Rain has also left the plants and trees gleaming, bursting in plenitude.

Natural irrigation in excess creates its own slow-rot, a sublime slime of wet decay and birth, profusion and irresistibility. Rain has this special seductive appeal — its innocuous wet, its piercing strength, its gentle drizzle-caresses, its ability to douse and arouse. The entire charm lies in its simplicity.

The September showers came too late, giving ample time for a prolonged drought. But when they eventually arrived, they brought with them the full fury of an unstoppered monsoon — the rain pelting down hard, cracking open newly laid tarmac, exposing the earth and the elements once again.

The pouring water persisted, overflowing until everything was affected — weak roofs, power lines, trees, un-warned shelters, people — almost everything.

After two weeks, the storm subsided — a war-struck wet wake — everything lay shattered in the aftermath, hungry, heavy, and low like polluted clouds of mist over a submerged *mofussil* that was trying to breathe and periscope back to life.

But here, the arteries are severed too severely to recoup its strength anytime soon. With or without water, in flood or drought, the existence here remains unchanged.

꒰

3

Pandemic

LOVE IN THE TIME OF CORONA

there is nothing you don't devour
— PABLO NERUDA

In the mirror it's Sunday
in the dream there will be sleeping,
the mouth speaks the truth.
— PAUL CELAN, 'Corona'

ASTHMA

Inhaler held to my heart like a prayer —
coughing, winded — I gasp
— KATE FIRTH, 'Breath'

Wheeze whistles — piercing shrill pan-flute notes.
Turbine blades slice my lung's trachea, bronchi.

Coarse sandpaper grate, ultrasonic — air-currents
struggle to clear my windpipe's fatigued length.

My eyes blood-shot in acrid distress —
dust mite, cat hair, particulates draw toxic tears.

My rib-cage tangled in its brutalist architecture —
my heaving chest tries its best to clear the choke.

It is such a struggle in this damp dust-weed air —
merely nonchalant breathing is a blessing.

ॐ

LOVE IN THE TIME OF CORONA

I don't believe in God, but I'm afraid of Him.
— GABRIEL GARCÍA MÁRQUEZ, *Love in the Time of Cholera*

In the dark times, will there also be singing?
Yes, there will also be singing. About the dark times.
— BERTOLT BRECHT

1.

Faint indigo tints in the greys of your hair
   evoke memory — Krishna's love for Radha,

its perennial longevity, its sustained mythology,
   its blue-bathed lore — such are life's enduring

parallels. Fourteen years — yet my heart flutters
   infatuated like first love. My hands fidgety,

palms sweaty, pulse too fast to pick —
   I am not allowed to touch your face.

Cyber-flurry emoji-love cannot assuage fears —
   or corona's comatose cries. *I don't believe in God.*

2.

In thousands, migrant workers march home —
   hungry footsteps on empty highways

accentuate an irony — 'social distancing',
   a privilege only powerful can afford.

Cretins spray bleach on unprotected poor, clap,
   bang plates, ring bells, blow conches, light fires

to rid the voodoo — karuna's karma, infected.
   Mood-swings in sanitised quarantine — self-

isolation, imposed — uncontained virus, viral.
   When shall we sing our dream's epiphanies?

3.

City weather fluctuates promiscuously
   mapping temperature's bipolar graph —

tropic's air-conditioner chill, winter's
   unseasonal hailstorm, sky's pink-blue spring.

Blue-grey will moult into salt-and-pepper,
   ash-grey to silver-white, then to aged-white.

My lungs heave, slow-grating metallic-crackles
   struggle to escape the filigreed windpipes —

I persist in my prayers. *I'm afraid of Him.*
   Hope, heed, heal — our song, in present tense.

54

OBITUARY

They were not simply names on a list.
  They were us.
— *The New York Times*

Death knells peal, numbers multiply,
  virus ravages us, one by one.

Newspaper columns loom, unsteady
  ghostly apparitions on broadsheets —

name, age, date of death —
  tall epitaphs in fine print.

Ink spills, bleeds dark — newsprint
  blotting out our wheezing breath.

No amount of hygiene-ritual
  enables our lungs to resuscitate.

Our lives — micro point-size fonts
  on an ever inflating pandemic list —

black specks, fugitive lonely numbers —
  the deceased, on an official roster.

Another sick, another dying,
  another dead — yes, *they were us*.

HOPE: LIGHT LEAKS

for Kwame Dawes

Darkness cannot drive out darkness.
Only light can do that.
— MARTIN LUTHER KING, Jr.

Late at night, light leaks — spilling
   beyond the door's rectangle edge —

a cleaving schism, its shape —
   a partial crucifix, a new resurrection.

Light's plane waxes, wanes —
   viral load expands, contracts.

Photons spill, conduction sparks —
   light slow removes cataract's veil.

In this *black*ness, *lives matter.*

SPEAKING IN SILENCE

after Fiona Sampson

Breathtaking weather surrounds us in these dark times.
I find calm in *Come Down* — misreading your book

title as 'calm down' — as if I am seeking balance for us
and others we love. Island climate can be promiscuous.

Indian songbirds outside my study tweet, pigs on
your English downs grunt — texting a common tongue.

> *come down / as everything breaks off / mid-story ...*
> *tractor tracks / a bucket at a gate / traces of the ones*
> *who left just this morning / centuries ago*

It was centuries ago, yet I know this place well —
we have walked together in this slurry and squelch.

In the coppice, I picked a driftwood piece —
sculpt-etched by wind-water — a palaeolithic

talisman I left on your rustic kitchen window. Perhaps
it lies there still — exactly there, on the sunlit sill.

> *and water shakes / old terrors / loose ... you could take /*
> *this with you / your whole life ‖ now everything / begins*
> *to move / and everything stays / where it is*

We speak in poetic phrases, punctuated by dactyls
and trochees, inundating line-breaks with half-rhymes —

this is the only language left, our private renga —
ancient codes dictating our syntax, not our accent.

As the world pandemically wrestles with dry heat
of disease and pestilence — profiteers pry, pilfer.

*dry season / in the heart / you have to pray /
although you can't /   but still the valves /
of the magnolia / wrench / themselves upward*

Marigold-magnolia will bloom, nature will dance
perennially — fine-tuned in its horror and beauty.

It is we who do not heed its signs, understand
its corpuscle-conduct. We have long lived in lockdown —

'social distance' in solitary silos — mutating metaphors
spilling everywhere, defying state and statelessness.

*flowers crowd / out of branches / that are holding / dark air up /
everything is / and knows it is — / wild equinox ||
everything to come / hides its face / among the shaking tongues*

I am certain we will continue walking, together and alone,
now and in the life-after — one's only guaranteed a lifetime

at most. Our silent speech stretches — like white,
its colours radiating beyond its spectrum-bandwidth,

beyond its infrared-ultraviolet — beyond infinite frequencies.
Whether it is chakra's sacred science, or just human belief —

*time folds / into another century / where you come walking /
down with them / into the future / they won't arrive at ||
such a tender rite / it is that brings us home / to the light*

that vanishes / and returns. Shall we return to aleph's source,
to oxhead's hieroglyph, to fountainhead's genesis —

to *a child / on a rope swing*? Are we on a robotic treadmill,
on a journey mapped out for us? Do we script our destiny

unlike nature's rolling hill-rains, unlike heat-dust-pestilence?
For now, let us *come down* for calm — to pause, reflect, love.

> *you were always here / in the body's / forethought*
> *in its heft //   heat and juice / in the smile /*
> *of a stranger / who will never / speak your name*

🌱

[NOTE: The lines in italics are taken from Fiona Sampson's book,
Come Down (Corsair, 2020)

OBITUARY 2: NINE PINS

One by one they are dropping dead
    at the rate of a heart beat.
Nine people I've lost in less than a week —
    Mangalesh, Mahmood, Asif, Astad,
Sunil, Vikram, Rachel, Karuna, Wahida —
    named and nameless.
Italicised epitaphs in multilingual script —
    so many that mere counting
leads to asthmatic laboured wheezing.
    This isn't a macabre game of nine pins —
but living souls pinned to the gallows
    prematurely. Covid's curse — R.I.P.

Covid 19
  buried deep inside,
protein cells create havoc —
  virus's vatic

꒳

Ventilator
  ventilators can't
arrest death-graphs anymore —
  futile pumping, still

꒳

Vaccine 1
  sick now sicker, dead —
hospital's revolving doors —
  cure still awaited

꒳

Sanitiser
  endless handwashing,
sanitisers, gloves, masks — a
  new apocalypse

꒳

Lexicon
  new dictionary —
'social distancing', 'lock-
  down', 'isolation'

꒳

New Normal
  our 'new normal' —
quarantine during peacetime —
  viral war, rages

❦

Pandemic 1
  will we find a more
compassionate world, after
  this pandemic's death?

❦

Pandemic 2
  pestilence prevails —
politics pushing aside
  science, medicine

❦

Rose Petals
  fighter jets shower
flower-petals on the poor —
  why not food, money?

❦

Trump Foolery
  Trump foolery — bleach,
Lysol kills Covid 19!
  Why can't he take them?

❦

Mental Health
    lockdown's uneasy
solitude — turning into
    another disease

❧

Death
    dread of death, death of
loneliness — our choices
    out of our hands

❧

Hunger
    migrants chew dry leaves
off the streets — no food, water —
    national disgrace

❧

Migrants
    walking the highways
hopelessly, towards fractured
    dreams, awaiting death

❧

SALINE DRIP

Sweat beads trickle from my forehead
    threading through my dense eyebrows,
over the protective arch of lids and lashes

into my unsuspecting eyes. My vision
    is awash in a fuzzy saline glare,
its sting fiercer than the viral load.

Perspiration transforms into tear drops —
    such is the potency of salt water —
brackish, transparent, intimate, deathly.

❧

VACCINE 2
   haiku

   untested vaccine
jabs in to an unwilling
    arm — black, colour-blind.

❧

NEWSREEL

When evil-doing comes like falling rain, nobody calls out 'stop!' / When crimes begin to pile up they become invisible. When sufferings / become unendurable the cries are no longer heard. The cries, too, / fall like rain in summer.
— BERTOLT BRECHT, 'When Evil-Doing Comes Like Falling Rain'

9am: Morning News

At nearly 50°C, you do not need a pandemic to remind you of human agony and grief — you inhabit one.

Daily death toll rises, coronavirus continues to inflate, infect. Enforced lockdown, inhuman laws — weary migrants, hungry, die on the highways.

2pm: Afternoon Bulletin

Cyclone Amphan razes the eastern seaboard — devastation — no power, flood, disease, millions homeless, millions quarantined.

Trees uprooted, electric poles down, cars submerged, shanties washed away. In Kolkata's College Street, soaked pages of books float in anguish.

Politicians continue their sloganeering — false promises for election, an eye for profit. *Evil-doing comes like falling rain, … crimes begin to pile up — they become invisible.*

Over the deserts of Rajasthan, swarms of ravaging locusts — destroy crops, livelihood — billions more heading towards the capital city.

10pm: Nightly News Prime Time

After the storm, when the rains diminish to a soft drizzle, the showers on the big-leafed trees echo and murmur. Or is it the distant sound of approaching pestilence swarms?

Next Day: Re-Broadcast

The newflash flashes again — repeating itself, primed for prime time.

When sufferings / become unendurable the cries are no longer heard. The cries, too, / fall like rain in summer.

BLACK BOX: ETYMOLOGY OF A CRISIS
for Neelam Mansingh + Kabir, Angad & Rocky

1. BEFORE: *Talk, Trailer, Fore Play*

"Prisoners of drops of water, we are nothing but perpetual animals."
— ANDRÉ BRETON & PHILIPPE SOUPAULT, *The Magnetic Fields*

A high-voltage swiveling lighthouse beam blinds us in this controlled darkness — Virginia Woolf or Robert Eggers are not present here to write their scripts. Shrill echoey electromagnetic sounds shriek, deafening our wavering eardrums.

Behind a lit translucent cloth-screen, a man in a wood-chopping motion wields his axe. His long hair glimmering halo-like — a chiaroscuro. He shines his shoes, breaks bread. He rummages through a box to a find a length of gauze to bandage his eyes, his mouth. His nose, stuck-shut by black tape.

On his bare body, he places fresh flowers on his skin, every hair follicle marking its petaline scent on a digital oximeter — measuring his pulse-beat, heart-rate, his blood oxygen levels — new-fangled indices of health, trendy obsessions of these pandemic times.

On an empty chair, sits a bodyless form — legs crossed, no spine, a jacket hung on the chair's frame, a spotlight glaring on it. This light moves, trailing a pair of footsteps, following electrical wires to a set of old switches blackened since war-torn blackout days.

A female form scribbles text on a notebook — *all work and no play* — in robotic repetition. Metronomic words leading to more words in silence — but speech cannot be silenced under any circumstance.

There is agreement and contradiction in this duality — a bipolar tension of ego/alter-ego, of fulfillment and vacuity in our unstable psyche. The graph is not constant or regular like sine or cosine curves — the mathematical grid inexact, unsure and asymptomatic like the contagion surrounding us — as we try to resuscitate every molecule of breathable air under our masked pretences.

In parallell, a film unfolds in the black box — the eye of Kabir writes dohas on an old tin trunk, the couplets composed in cinematic frames, its edgy noir feel obliquely reminiscent of *Mehsampur*.

In *the company of dark matter*, I try to trace my steps of sanity in this thick heavy air as we sit at more than an arms-length fearing human touch and disease.

What convoluted times we live in now — where being inhuman is human, where free-thinking is dissent, where being democratic is anti-national. Even the 'black box' of a crashed airplane storing facts cannot reveal the facts — everything in done in secrecy, everything is subterfuge to continue oppressing the subaltern, everything is about power or the lack thereof.

ॐ

2. AFTER: *Play, Black Box*

It was the best of times, it was the worst of times.
— CHARLES DICKENS, *A Tale of Two Cities*

Under a conical thatched roof held up by bamboo armatures, the *mise-en-scene* —

four tin trunks painted black containing personal and household items, two off-white gold-bordered cloth curtains, a metal kettle, a large white shallow tray to hold water from spilling out, a metal black chair, a white bicycle, white flowers, black electrical tape, two long bamboo poles, a pair of shoes, two empty transparent polythene bags — all framed by three bamboo poles set up as a goalpost, or a proscenium marking out territory to contain spillage of any narrative beyond silence.

The theatre walls are painted matte black, the floors tiled in clay terracotta. There are three lights that hang from the ceiling, a whirring fan, two spotlights, and old-fashioned wooden switchboards with clunky round black switch fixtures.

The lead actor, a soloist dressed in black, lies askance on the carpet on the centre with a low wall of loosely stacked bricks forming a horseshoe enclosure.

It is silent here and our eyes are led by lights that train our sight to follow a story. It is a gaze that looks outward and inward. Words are minimal, metronomic, repetitive like a refrain

mouthed by an invisible chorus. But there is
no ensemble cast or musicians.

❦

A white bicycle stands at an angle, alone —
white flowers lie scattered, upturned on the
floor. Piece by piece, petal by petal, I pick
them up — stick them onto my bare body, on
my eyes, nose, mouth.

I get onto the cycle — slowly start pedaling,
circling the outer periphery of the brick
enclosure, marking my tracks. I gather pace
and more pace, circling round and round
at breakneck speed. I disembark— start
unpeeling the flowers off my body ... and
start running, circling round and round like a
falcon.

*Turning and turning in the widening gyre / The
falcon cannot hear the falconer; / Things fall apart;
the centre cannot hold; / Mere anarchy is loosed upon
the world, ...*

❦

I sit on the loose bricks that form an enclosure,
a geometric U-shape. I sanitise my hands, look
at my hands closely, surveying my destiny
— wear latex gloves, put a pair of shoes in a
transparent polythene bag — and say: *"Yeh sab
ko chahiye nahi!" "No one wants these!"*

Dateline: March 25, 2020. I start counting
from 1 to 2 to 3 to ... bang a steel plate with

a stick — keeping up the beating until the cacophony is no longer discordant.

❧

The country's future seems futile. No earnings, no hope: *"I want to go back to my home"* ... 485 kilometres ... I start to count down the kilometres as I drearily trudge along the highways and on rail tracks with a trunk on my back like a homeless migrant. All transportation is shut down, everything immobilised — our mobility too is immobile — *"I can't go, can't reach home!"*

I am craving for a home-cooked meal, a simple meal — but instead all I have is a stale sandwich donated by someone who took pity on me.

❧

Sanity, insanity, sanitised — I give my half-eaten sandwich to a stray animal and insects. Both animal and human reduced to one, on our knees, by the powers that be. Everything is shut — door, window, sky, auditorium, stage, audience — no one is spared.

What does an artist do? Storytelling, *dastangoi* — stage, kings, *ghungroos* — story of a sparrow, wise folk tales — anything, any tales to keep our imagination intact, alive.

❧

The overhead lights dangle precariously on wires that might short-circuit any moment — like pendulum clocks, they waver, counting down time. *Ghungroos* become an instrument to auto-tune dissonance in place. A moth sits on the white screen, its wings wingless, awaiting flight.

Sparrow speaks to an ant, the ant to an elephant — everything is a deal, deal without a deal, deal within a deal —

🦗

Dateline: April 24, 2020. Body, corpse. My cadaver tries to sit up, ascending with the help of crutches. I spot a squirrel. I walk, walk, walk. *"Let's go for walk, … baby, let's talk …".*

Two transparent polythene bags. I fill water in them using a kettle, and then tie them on the bamboo crosspiece. I pierce the bag carefully with multiple holes. The piercings induce rain. Rain is the only hope.

🦗

Chair with a wet cloth, bodyless, waterless, hopeless — nullity — everything, all life-source is pushed away —

But still one is hopeful — writing on soil — scripts of hope, future.

"Hello, hello …!" — just for a moment I am reminded of Pink Floyd. *"Hello! Is there anyone out there?"*

🦗

Four tin trunks. Two of them the same size. Ideal furniture.

He opens a trunk that contains stories, cloth, a blue floral-printed woman's top — vesture of memory, hope.

Only in dreams, is there hope — hope of embrace, humanity, scent of my beloved's garment —

"Where are you?" I can't hear you, touch or feel you. All senses have evaporated. I have nothing. I have everything. All my mere belongings in a trunk. I stack the trunks up in ascending order, and open the smallest one on the top. On the obverse of its lid is pasted a 10-digit number. Can I call for help? Or is it just a missed call?

Om, om, om, … breathing – *pranayam*. Back to the beginning. Black box.

We look at the world once, in childhood.
The rest is memory.
— LOUISE GLÜCK

It is very late at night and I am knackered. Yet,
I cannot sleep. All night I dream in fragmented
images. Memory plays tricks with my mind.
Her story, his story, my story, history — all
collude and conflate.

The trailer I saw before was only a glimpse.
The film is still being cut. We might yet
change the narrative. But do we have control
over our own destiny or karma's fate? Jump-
cut, dissolve, fade. The parallel sprockets of
analogue film-reel struggle to run smoothly on
the spool.

It is all digital now — memory is not an
issue, megabytes abound in tiny microchips.
Yet it is all about memory — real, virtual —
inscriptions on epitaphs, coded hieroglyphs,
ink, text.

Will. Our will. A will on a parchment that was
never written. Will to live. What will it be? *"Is
there anybody out there?"*

ॐ

QUARANTINE

I can hear — my heartbeat quietly
sing — the metronomic constant of an
antique clock ticking — wind's rolling
rustle outside — chirping birds, mothers
feeding fledglings. I witness — the
changing colour of leaves and skies —
the secret night-whisper of the stars.

In the company of myself, I reflect.
It is time to call family, a neighbour,
a neglected friend — time to read,
rejuvenate, revive — rekindle *love's labour
lost* — time to savour life's little joys.

These have been around, but we didn't
make time — now we have time, we
complain about quarantine.

4

Contagion

CORONA RED

How far from the beginning are we?
When did we first start to feel the wound?
— SUSAN SONTAG, *Unguided Tour*

IMPLOSION

A storm is raging inside my ribcage
— electric sparks short-circuiting my
bronchial tubes. It feels as if my lungs are
imploding into themselves with a severe
lack of oxygen supply. Constant high
frequency whistling sounds of wheezing
pierce my eardrum. I cannot sleep through
this night's cacophonic cyclone even though
all appears calm and still.

On my bedside table, even the electric
bulb under the lamp's hood cannot hold
its wattage steady with all the fluctuations
inside me — mirroring only mildly, the
tsunami inside.

I need to call an ambulance, but I hesitate.
More eucalyptus steam inhalation, Ventolin
sprays, mixed concoction of ginger, black
pepper, turmeric and organic honey,
provide only a temporary respite.

A looped orchestral score of unbalanced
shrill echoes inside my chest making
Stravinsky's 'Rites of Spring' sound
melodious. I cannot breathe — I desperately
need air.

᰿

'O' ZONE

The spray of scented chill pierces my lungs first, then comes the slow desperate heaving, the grinding spasm splaying, trying to centrifuge stubborn coves of mucous — whose greenish-yellow viscosity remains more deceptive than quicksand's subtle death trap.

My face — confined in the transparency of plastic, frosted glass and thin air — regains for a moment the normalcy of breathing. It is a brief magical world. The oxygen in my blood is in short supply. I feel each and every electron's charge, spurring my senses.

Dizzy in aerosol hope, I try to free myself of the medicated mask, but the frozen rain that batters my face reminds me of the tentativeness of living. As I survive on borrowed air, I'm grateful to the equation of science, its man-made safety, its curious balance that adds that precious molecule to create the sanctity of 'O_3' — the holy Brahmanical triad — and the triumph of its peculiar numeric subscript.

My breathing is temporarily back now — electrolysed, perfectly pitched and nebulized — as narrow transparent tubes feed dreams into my wide opaque palate.

The sun's edges are dark, so are my heart's. No amount of air will light them up.

Cold blast from an electric vent bites
my skin — this comfortable discomfort,
prickling my pores bathed in an acrid glaze,
transforms to frozen gold-salt.

Attaining instant freezing points might be
a rare marvel of science; I like this hellishly
good blast that shakes all the embedded
molecules in my bones —

bones that are parched in heat, turn to
skeletal icicles — a beautiful ballerina-
geography of stalactites and stalagmites —
each needle-end points towards the other

like the two longing fingertips in
Michelangelo's painting at the Sistine
Chapel — desiring a touch.

🦅

FEVER PITCH

I wake up cold, I who
Prospered through dreams of heat
Wake to their residue,
Sweat, and a clinging sheet.
  — THOM GUNN, 'The Man with Night Sweats'

The seductiveness of a slim tall transparent glass tube — the curved silver juices it contains — is such that it makes me forget the news of the birth of a new child. Human life and inert chemical life compete in insidious ways, the same way fact and fiction do, as do desire and disgust, illness and passion.

Like an aria, it is a curious melody, as distinct from harmony — a solo part in a cantata or opera. Its inherent nobility and splendour, its treble and bass create an enigma of its own private architecture.

The mercury in the thermometer rises, gradually and numerically, to a height where human equilibrium can just about balance itself. I stand at its base. The glass chamber rising many storeys above me holds a reservoir of finely granulated liquid that changes its silvery-grey shade in the fading light. Above that, a constriction, then a towering shot of fine tubular glass hoping to reach a degree of sanity at the cost of human heat.

Summer is already approaching outside; my
body sweats gently in appreciation. The
heat worn by my skin's surface is nowhere
near the heat that is slowly welling up
inside me. It takes the lightest of touches,
a feather-swivel for it to shoot up the scale.
But, at the moment, all is calm as the storm
gathers pace.

I am dying for the monsoon rains —
but I am caught. Trapped in the wrong
longitudes, these wet dreams are dreams
that will have to remain un-soaked. The
hair on the surface of my skin itches to
raise its hood to attract any pheromone in
sight. There is a magnetic lull and hush, a
loud silent sound of breathing, in different
voices.

Platoons of clouds clash softly without
any hint of thunder. There are electrical
impulses that are waiting, poised to
spark. But the perfect noiseless moment
is what everyone is waiting for. Only the
obtuseness of instrumentation can clarify
that, but that would be too intrusive.

The mercury shows its first sign of life —
a little trickle, then a tremor, then a
surreptitious U-turn past the erectile
crystal-tissue. Thereafter, complete
freedom. It is at this point that the human's
heartstrings and the chemical's soul marry
perfectly. Each follows the other's actions,

responding on a natural impulse, like the soothing scratchy sound of ice severely eroding under a ballerina's silver skates. Metal matches metal, breath matches breath, glass matches ice, freezing the heat itself.

I sit — serenely delirious — on the convex tip of the mercury's crest. All around me is vacuum — and beyond that glass — and beyond that a semblance of life and world. Here the vagaries of temperature do not seem to matter — a sanitized skyscraper holding the elements of inertia and energy. Here I feel particularly buoyant, not because of anti-gravity, but at the hint of rising temperature.

This is the third thermometer I have bought in a day, and yet I cannot trust it. Twice before, the reading shot out beyond the graduated scale itself, hinting either I was heated to the point of insanity or it was a case of the glass's own neutral impotence.

This time I am determined to get to the heart of it, inside its very core, whatever the consequence. However, when one is caught in the process of creating a grand score, it does not matter what the root causes are. Genesis, like the Christian one, should remain a Buddhist mystery — then all religions can command the private power of the elements themselves.

Molten silver — boiled, cooled, boiled, cooled, boiled, then caressed variously over skin — finds an intimate space that intersects the point of heat — glows, dense and quiet. One knows the gravity of such events, but not their intimacy, not their relationship with follicles that create their own forest fires with their own human climatic changes.

It is these alterations that marry physics, chemistry, biology and mathematics — there is hope in all these — just like the sine curve's elasticity and predictability, the graph's nodes are stretched straight on the X-axis, the subjects collude to a point of nullity. At the point of birth, there is the death of the womb itself, but one lives — so there may be hope.

It is at such interstices that art and passion find their true shape. The unknown boiling and freezing points that I hide within myself provide the ultimate enigma that even the most specialized doctors and architects find hard to map. My body is a terrain that defies the contour of safe plotting — indices like Celsius, Fahrenheit, torque are all inadequate — just as bone marrow count, triglyceride, HDL, LDL do not form pretty, explainable equations.

Amid this oratorio, the cold tactility of a three-faced glass case, its triadic ancient constancy, its contained columned virility, provides comfort to my talisman. Sometimes even the most brittle seems to find some soft shape for hope. Silicates form so many forms — but what I like most is their stubborn transparency, their supine pirouettes, like the vicissitudes of mercury — like breathing itself — at least until they last.

HEAVY WATER

You were my death:
You i could hold
When all fell away from me.
— PAUL CELAN, 'You Were My Death'

There is something deeply arthritic about water
and pain, the way water seeps into unexpected
fissures in bones, the way it conducts pain
itself — operatically, electrically.

This morning I woke up, as I usually do, in
pain. It was a new sort of pain, a pain that I
had not encountered before, so I didn't know
how immediately to respond or manage it. All
this while, I had sorted and filed each type of
pain into neat bearable files, each with their
possible recourse to relief, albeit temporary.

It had rained all night, and this morning
it continued without any relief. The sound
of persistent rain once provided calm —
but all this water sound, with its chaotic
decibels, were annoying my breathing,
heartbeat and sight.

Whether my sight was blurring due to water
battering my retina's windscreen or whether
it was triggered by the slow accumulation of
pain in my heart was difficult to measure or
analyse. Only intensity and volume mattered
— cubic litres, millilitres — almost any
equation with letters and numbers raised to the
power of three. Triadic superscripts — n^3 —

there lay some oblique clues, but perhaps only to the initiated or those who wished to be part of its intimacy.

The irony of intimacy is such that the closest in the family seem the furthest away. Their attempt to be interested, in spite of being uninterested, ultimately measures pain and its intensity. Intensity is a peculiar thing — its measurements are tactile and ephemeral, quantifiable and infinite. It is measurable, its heat and depth fathomable.

It is the ephemeral that is painful. Water creates all the confusion — its saltiness, its acridity, its mineralized purity, all compete in ways that chemical equations find hard to support or balance.

Families of electrons, protons and neutrons speed away, whirring in patterned loops, forgetting all the while that the heart of their orbit may actually feel and breathe. But in science, as in the ambitious ruthless route of success, there is no room for unscientific thought — as if science and the arts, coolness and emotionality were mutually incompatible or different from each other.

I am in pain, and I just want to cry, cry and cry — so that each searing cry can etch some fragment of a note, which has gone unnoticed, so that each measure of pain is no longer diluted for people who listen because they have to.

I wish to paint a canvas that invents new indices of pain and water, for anyone who wishes to listen and bear, for anyone who wishes to understand — not because they need to, sitting comfortably straitjacketed — but because they are moved by it. We need to be moved, moved by the finer chords of music and art, so that both electricity and opera can operate as they always did, in tandem.

But heavy heart, like heavy water, is difficult to dissolve — their melting and boiling points register unusual scales — scales that peal and peel, echo and layer, untying each and every fibre that breath requires in order to survive.

NIGHT WARD

What cannot be said will be wept.
— SAPPHO

The night ward's blue curtains that surround me drip colour and deceit — each and every pleated flute of cloth hiding some half-truths like the half-lives of atoms. Only here, the arithmetic surety of fission does not wish to match the nuclear chemistry of my blood's transfusion.

The night nurse peeps in to assure me that blue is not all black, that red is not grey, that the colour of my skin does not reflect the colour of my life. I wish I could agree with her consolations.

Yards of white and blue linen that wrap my slow generous chill, know the real secret of my floating corpuscles — the flotsam larvae, their ancient silk that gently threads my nearly finished mummy.

I want to be shattered like a dream
Such a loneliness that wants to die
— IFTIKHAR ARIF

Get your papers in order — choose
your inheritors fairly — with love, care.

Outline clearly — who gets what,
what they are required to execute.

Execution after your execution —
their inheritance, your legacy.

Thereafter, the phase of reflection —
call all who you wish to one last time,

forgive those who have wronged you,
smile, hug, and give gratitude.

Record everything in minute details —
leave no unresolved business or debts,

donate your organs, give to the needy,
veer on the side of being generous.

Then, the most difficult part —
how and where to die, what to wear.

Be tidy and smartly turned out—
there is no room for shabbiness here.

Of course, one would like it to be
swift and painless, without any show —

an elegant private ceremony for one,
a dream end, a perfect death.

ICARUS

The image of Icarus has been flying around in my head. I cannot get rid of it — I tried lopping off its ill-fated wings, persisted in pushing them close to the sun to burn them off, so that I could erase this myth forever. But myths like imagined truths and cold silence are very hard to shake off — they induce sweat, heat, palpitation, even illness.

I cannot sit still in this beautiful Goa's warm winter weather — so I go to the bedroom to lie down under the air-conditioner's induced cool — but the feverish heat will still not leave me. My mouth has new ulcers, lip sores bloated, sty in my eye engorged, and tears welling up. Nothing I do, stop these. The brine in my tears is so strong that I feel blinded by its corrosive salt.

I pray for Icarus to return to take me away on its imperfect wings until we both perish — since the myth's script willed it so. I could jump off this high-rise balcony on Teligaon Plateau, catch the ends of Icarus's feathers and somehow hold onto them. Come Icarus, come — take me away — let us burn together! Doing it alone is fearsome, heartbreaking. You have already done this before, its example mythologized — so why not another time for a fellow soul whose flight path is derailed, deranged.

I see you clearly, Icarus, circling over the seawater in Mormugao Bay. The inflamed orange sun will burn you — I know that from the tale already recorded. I see you struggling to keep afloat in the air — your wings slow-fluttering, unable to keep pace. Why don't you fly past me one last time so that I can jump onto your soft back — and we can then nose-dive into the craggy shoreline — where a graceful death awaits us — operatic, full of colour — amber-golden light mixed with fresh blood that spurts out when my skull smashes into the rocks at great speed. What a wondrous celebration of Holi it would be — full of ceremony and ritual — full of flight and dance — full of heavy-hearted fables.

And before I know it, before I realise that I missed catching your burning wings — I hear a hurtling collision. Human and bird flesh, skeletal cage, brain, imagination — annihilated in a serene implosion before impact, just short of meeting the rocks and water — our hearts punctured beautifully and symmetrically — a new myth, a poem, half finished, smashed to smithereens.

THE LEGACY OF BONES
for Adil Jussawalla

It's high time the stars were re-lit
— GUILLAUME APOLLINAIRE

I buried my body in the same soil where I had learnt to crawl. I waited until my skin decomposed so that I could rescue my bones to craft new implements to write with, anew. Imagine making bone-nibs of various sizes and intricate patterns that contain your own tissue and imprint. I waited, waited until the magic of metaphorphosis could take place. It didn't matter whether it was in my lifetime or not, clearly I still waited having performed my own burial. A fine anachronism — even Sophocles would be astonished, or the tales of Gilgamesh might have been realigned.

As I re-live, piecing my first alphabets together — elongated letters form arcs and loops — creating a score, a grand opera where bone nib-tips play a crucial part in the sonics of the composition. I am still tuning them in my mind as I wait with the dead — the dead to fill in the chorus, the dead to conduct the show. Whose imprimatur shall the music bear — what shall it be called? Fibula, femur, F-sharp — fine featured whispers layer its richness. Where is the ink, the ink familiar to every bone? Blood. There is no blood left now.

But air has sufficient magic left — its slipstream modulating a script that has not been written before, notations using my DNA to code the coda. The earth says she wants to name it — I say, say it aloud. She prefers a subtle sigh that comes with the quiet confidence of permanence. Gradually the aria begins — singing of the eternal purity of bone music.

It requires music for bones to patiently heal. It requires compassion to love selflessly. The buried vatic song starts to leak, leaving the legacy of our bones. It's high time our bones started to sing aloud. *It's high time the stars were re-lit.*

from an ongoing series, 'Small Tales from my Kitchen Table'
photograph by Dinesh Khanna

Two knives lie like a dismantled fallen
crucifix — is this a new metaphor of our
times? It is difficult to get a handle on
things — even with sanitised gloves I am
unable to chop clean every seed, every
rind's red heart.

Onion's oblique pungency may buy us
time's illusion, flatten the pandemic's curve
during isolation lockdown, calm corona's
death wish — but it cannot deceive the
cutting board's old seasoned wood which
remembers everything — the past, the
present, and what will be.

The circumference arc of a transparent
Petri dish like that of the sun or a flower's
pollen-packed centre, offers hope. I hope
that the edge of the knife's razor-blade can
cut through the unsuspecting errant garlic
clove, fumigate it, provide some safety to
all.

Even when religion's badges are dismantled,
its core universal song for humanity and
humanness remain. It is time to feast, share
bread and sing — or else, it might be too
late to catch the lyrics.

৺

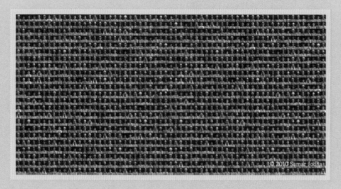

'3500 Portraits' (2004-10) of each man who helped built the world's tallest structure, Burj Khalifa. / Cast concrete wall 8'x40 (2.43 x 12.19 metre) by Samar S Jodha.

MAY DAY
for Samar S Jodha

'May Day, May Day' — is it a distress call, praise song, war cry, or just another day? Is it lip-service marking 'world workers day' — social media slogan without consequence?

We communicate in exclamations, emojis, emoticons — intricacies of words, beauty of language, their epistolary power, lost to texting.

Tomorrow we'll forget the migrants' faces, their lives balanced on destruction's edge — replace them with another face, another hash tag, another named day? It is not just any day — 'May Day' is every day.

🦢

Scar

The heating filament snaps —
     orange-white death —
electric rage.

Winters in Delhi,
         unpredictable
like its power supply —
     adulterous,
unreliable, fickle as weather.

Even vane's direction
meaningless,
     winds' changing moods —
unsteady, dubious.

Mistakenly at an old friend's
birthday gathering —
     a non-dinner invite
at a dinner do.
     Are friendships tested
in pandemic times?
Socially distant — alone
in a noisy crowd — I left early.

Weather these days,
     deeply polluted —
even masks can't prevent hurt
or lung damage or heart
            break.
Burnt-out bulb fragment
     falls on the cold floor,
charring the concrete —
     bold scars —
     that is how science
and practicality confront
human emotion.

ॐ

GHALIB IN THE TIME OF CRISIS

Ranj se bhū-gar hu'ā insāñ to miṭ jātā hai ran
mushkileñ mujh par paṛīñ itnī kih āsāñ ho ga'īñ

If a person becomes accustomed to grief, then grief is erased,
so many difficulties fell upon me, that they became easy.
— MIRZA GHALIB

In Chandni Chowk's narrow uneven lanes of Ballimaran, the ghost of Mirza Ghalib is real, even 300 years later. His apparition still resides in a half-restored home, propped up as museum in Old Delhi — where memory and memorabilia share a wall with a pigeonhole photocopy kiosk. Outside, the lanes are dug up — and after the rains you cannot avoid the muddy squelch.

A maze of electrical wires, cables, and telephone lines crisscross overhead linking adjacent buildings in a mesh-like parasol — their intricate weave, the structure pattern of a *sher*. Woven tightly, Ghalib's ghazals remain intact, preserved and passed on as couplets and enduring songs — his wise eyes in the portraits still searching, looking into the future for peace to arrive.

A pair of perfectly scanned amatory couplets — *sher/bayt* — looking longingly at each other on facing pages — maintaining distance — cannot mate. In this silent formal dance of *matla, radeef, qaafiyaa, makta, beher* — there is balletic mirroring — mumurations of measure and metre, love and pain, reality and lie.

In the adjacent, fluorescent-lit, photocopy shop, there is constant whir and churn — as sheet after sheet passes over the machine's glass flatbed metronomically, interspersed with blinding flashes. Photocopy's banal replication and mindless mechanisation are apt metaphors of our times — digitise everything as world fodder, the web converting all into bit-mapped clones, a spineless misguided chorus.

Social media, judge-and-jury — Internet debris, a permanent scar. Fake news like the coronavirus replicates at an alarming level. Multiple pandemics prevail — politics of profit and power, mistrust and misuse. In these Seamus Heaney lines, there is hope: "If we winter this one out, we can summer anywhere."

Our life on this earth is miniscule, just one brief instant. In the cloistered safety of my study I dream and create — unformed, embryonic dreams, waiting to be shaped and crafted. Etched in stylised *nastaliq* loops, slanted lithe ascenders and descenders dance — now, even tarnished gold will glow anew.

بـلآغ مزبس ےس راویـد و رد ے مہ امر گا
ے مہ یئآ راـمب ریم ریم ر ھگ روا ریہ ریم ریابای ب مہ

Greenery is growing from walls and doors, Ghalib
I am in a desert and my house blooms as if it is spring

ॐ

5

Atmosphere

SKYSCAPES

Green was the silence, wet was the light,
the month of June trembled like a butterfly ... .
— PABLO NERUDA

SKYSCAPES
for Rita & Aditya Arya

In the lockdown period, I would spend two to three hours in the evening on my terrace — reading, walking, watching, calling out to neighbours, eavesdropping on birds, "listening to the stars", and photographing skyscapes.

I took to capturing the skies from exactly the same vantage point on my terrace day after day — and the selection of photographs here will give you an idea of the varied vivacity of the ever-changing canvas.

— S.S., from the 'Prologue'

late at night, light leaks — spilling …
a cleaving shift

ink spills, bleeds dark — newsprint
blotting out our wheezing breath.

blue-grey will moult into salt-and-pepper
ash-grey to silver-white, then to aged-white.

light's plane waxes, wanes —
viral load expands, contracts.

flames forming huge flares,
fragmented waves of golden-amber spark,
electrifying helical fire-flurries —

tree silhouette, fixed —
sky-air-hue, ever shifting —
kaleidoscope frame.

the ray's glare / splits open their perfect coronas —
pollen shower-burst, an ochre-flare,

amaltas / ∂rip ochre at 48°C,
∂renche∂ in yolky heat. //
Hotter the heat, more
incan∂escent its colour.

listen to the stars — far, flung apart —
elsewhere, nowhere, everywhere.

6

Holocene

GEOGRAPHIES

Oh, science, with your tricks and alchemies,
chain the glacier with sun and wind and tide,
rebuild the gates of ice, halt melt and slide,
freeze the seas, stay the flow and the flux
for footfall of polar bear and Arctic fox.
— GILLIAN CLARKE, 'Glacier'

The clouds pressed closer, blotting out the daylight.
The mountain loomed darker. It really did look like snow.
— FIONA ROBERTSON, 'A Shift in the Ice'

ABANDONED GODS

Sindoor-haldi smeared
  abandoned village gods —

revered by rural hands —
  come alive momentarily.

Hibiscus, rose, marigold —
  match fervour and devotion,

lending broken moss-laden
  weather-eroded deities,

veneer of a fleeting new life.
  It does not matter

if organic colours are erased,
  wiped out by errant rains —

or flower-petals wither
  in global warming's heat.

In it's temporary finery,
  at least the moment is living.

ॐ

Maihar, Madhya Pradesh

DRIFTWOOD

for Derek Walcott & Sigrid Nama

At the end of this sentence, rain will begin.
— D.W., 'Archipelagos', Map of the New World

1.

Part of the bannister-railing is absent
in spite of its strong metal-rivet moorings.

Termite-eaten, consumed by the sea,
I can see its woody skeleton float faraway

among the surf, its salt-scarred coat
tossing and struggling to keep afloat

against the waves' incessant lashing.
There is music in its disappearance —

a buoyant symphony,
note-strokes resurrecting life,

a new story — history restored
by resilient fingers of a master artist.

Wheelchair and weak legs
are inconsequential impediments —

his mind sparking with electric edge,
whiplash wit at its most acerbic.

There is generosity for family, friends —
those who are gone, and remain —

and thirty new poems,
an intricate magic of ekphrastic love.

2.

In the front garden facing the same sea
with Pigeon Island on the horizon's left,

lies a cluster of wind-eroded oval rocks —
their shapes mimic a lost egret's nest

or a ballerina's curved arch —
a stone-memorial for a close friend.

3.

The driftwood is now out of sight —
part of his house donated to the sea —

in gratitude the sea sings
a raucous song,

folded cumulonimbus clouds echo
in synchronicity — a soundscape

absorbing his commandment:
At the end of this sentence, rain will begin.

꙳

Castries, St Lucia

PERISSA BEACH: BARE FACTS

0.
Hierarchy
    of maritime geography,
   edicts ordained by Grecian goddesses —

1.
Rocks —
    jagged dangers, slippery,
   lurk innocuously under shallow shores.

2.
Stones —
    time-eroded shapes, smoothened
   flat and oval.

3.
Coarse sand —
    slow churning, continuous grinding,
   bleed human feet's foreskin.

4.
Sand —
    deep black, raw volcanic black —
   subtle white-greys, subdued.

5.
Fine black sand —
    last layer, super-smooth velvet —
   fine-grain soft-powder, black silk.

🦋

Santorini, Greece

AKROTIRI

Between Perivolos, Vlichada and Mavro Vouno —
beach sand changes from black to red to white to black.

Lime-ferric colour transformations — calibrated
chemistry of livid lava, physics' precise refraction.

Four centuries ago, Akrotiri's ancient site fell
grandly to volcanic death, victim of several quakes —

buildings, frescoes, ceramics buried under ash.
Excavations reveal — sand-soil compacted mineral

paintings — rainbow reserved normally for the skies.

Santorini, Greece

WITHERSTONE

for Fiona Sampson & Peter Salmon

The deaf don't believe in silence.
Silence is the invention of the hearing.
— ILYA KAMINSKY, *Deaf Republic*

1.

In bucolic,   translucent silence, I overhear:
stone-slates aligning themselves
                        in tiered mosaics —
floor's varying heights competing
                with rural mud-grass gradient —
sheep excitedly running downslope
            to greet their master, seeking salmon —
a quartet of chickens in an open shed
    pursing their rear, revealing pink eggshells —
and Wye waters creating an unintentional arc,
                        verge of an immaculate oxbow.

2.

Sam[p]son and Delilah, Church and State
            jostle, trying to carve their own space —
a map whose coordinates appear flawed —
                    fragmented like old Balkan fissures,
like Brexit's comedic miscalculations,
            like pandemic's political mismanagement.

Tudor stone's past — like a misleading folly —
    gradually withers away history and time
            as erosion's song-cycle prepares for a coda.
A perfectly-pitched aria or cantata
    calibrates its modulation in this wet heavy air —
the frequencies unsure,
            like a directionless weather vane.

3.

Magnolia's magnificence in the front garden,
     its regalia in temporary full glory —
before the snow-smitten air bites through its tree-bones.

In the large glass-paned sunroom,
     a long red cylindrical punching bag hangs listlessly
       waiting for an uppercut to deflect
a dangling modifier — a poet's primal prerogative.

4.

A red metal kettle in the kitchen, excited by heat,
whistles like an old steam engine on a disused rail track —
     brewing rose, green and white infusions.
As I set aside the coffee story,
the large black cooking range
     mirrors an age-blackened timber lintel.
Therein lie unrevealed, unsaid stories —
     stored within chamfered beam's wood cracks.

5.

     Wireless signal, desperately elusive here —
the valley's *rough music*
     rearranging the air-waves' diatonic notes —
its *common prayer* bridging
     the geographical *distance between us*
      in this *limestone country*.

6.

How topography fine tunes our sensibility,
     landscape reshapes our psyche —
how everyday banalities of potatoes, animal farm,
     persistent rain can soothe our senses to calm —

SUDEEP SEN

how simplest of neighbourly gestures
                    cements communal intimacies,
                              reorienting our DNA.

7.

Morse code conveyed in silence —
            skyscape, ever changing,
                    planktons floating on unreliable waves —
their dramatic formulations,
            shape-shifting cumuli,
                        thermal up-draughts
matching   a local brook's innocent eddies.

8.

At an abandoned countryside churchyard,
            I pause at each gravestone
to decipher its ornate genealogical etchings —
their looped serifs hold still the heartbeat of many lifetimes.

It's the kind of clock I want to measure time by —
            time that depends
      on the company of those who care —
                  time minutely layered
on this open windblown Herefordshire terrain —
                        an expansive canvas roll.

9.

Traversing a four-acre fenced land in borrowed Wellies,
            my pugmarks leave a foreign imprint on this soil.
I find among the muddy squelch,
            a piece of dead bark. Its smooth weatherworn
seductive shape reminds me of an ancient whale,
            its striated sanded-down skin bearing a script
            left undeciphered until now.
I am tempted to decode enjambment's mystery,
                        but I resist.

10.

Inside Witherstone,
the well-worn kettle-nozzle tweets again,
          a trio of iPhones peal their pedestrian pings —
I choose not to hear this uncoordinated medley.

                    In my imagined silence,
ceramic cup-stains graph every minute detail,
          letterform's each ascender and descender ||
                    as I drink my infinite cups
of bergamot oil infused tea   without haste —
                    slow-staring at the sky's ash-rose stories

ॐ

Herefordshire

Undercurrents: 20 Lake Haiku

for Janet Pierce

1.

water's mirrored sheen —
    swans skate, gliding on its skin —
marine undertow

boat-house wood walls creak —
    barque, oars, rowers — now absent —
memories, slow-leak

clouds lie on water,
    photo-perfect and stone still —
a transient frame

fish come up for air
    glimpsing skyscape, a moment —
refracted vision

fall-leaves coat the lake
    water soaking elements —
demersal debris

a shallow swimmer —
    skin and water burning fat —
fluid aerobics

ice-cold swim — healing —
    water's conductive powers —
sinews assuaged

2.

the dark waters, suck
   me in their undercurrent —
the ship is sinking —

volcanic rumblings
   destabilize my moorings —
the earth is shifting

lake's pristine waters
   infected by human greed —
pollution of soul

whispered murmurings
   eavesdrop, plot, hatch, weave — churning
carefully — back-stabbed

despite lake's leaden
   waters, irony's missing —
misplaced, forked values

lake's blue-black ink
   runs deep, piercing sinews —
leaving scars, unseen

3.

blue and green merge — still
   waters calling cormorants' —
pure white, refraction —

freezing waters thaw
   an artist's skin — epsom salt
hot-baths purifies —

clear sky, rain, slurry —
   lake's blue, trees' reflected greens —
soulmate's oxygen —

blue-greens change slowly —
   orange, brown, yellow, light up —
autumn colours bloom —

geese squeak, cormorants
   dive, fish summersault over —
invisible waves —

gentle breeze wavers —
   uneven reeds circumscribe
lake's surface tension —

a lone swimmer dives
   into the pristine sheet of
water — liquid graph

𝔶

Annaghmakerrig, Ireland

DISEMBODIED 2: LES VOYAGEURS

for Bruno Catalano

To understand yourself, you must create a mirror
that reflects accurately what you are ....
Only in the understanding of what is,
is there freedom from what is.
— J KRISHNAMURTI

Bronze humanforms sculpted, then parts deleted —
        as if eroded by poisoned weather, eaten away
by civilisational time —
                corrosion, corruption, callousness.
Power, strength, gravitas residing in metal's absence.

Men-women, old-young, statuesque —
                holding their lives in briefcases —
        incomplete travellers,
Marseilles les voyageurs, parts of their bodies
                        missing —
surreal —   steadfast, anchored.

Engineered within their histories
                of migration, travel — over land, by sea —
coping with life's mechanised emptiness.

Artform's negative space or positive? What are we to see?
        Have these voyagers left something behind,
or are they yearning
                to complete the incompleteness
in their lives?
        They are still looking —
                as am I,   searching within.

꒒

Marseilles, France

SUDEEP SEN

DISEMBODIED 3: WITHIN

for Aditi Mangaldas:

You emerge — from within darkness, your face
                        sliding into light —
you squirm virus-like in a womb,
draped blood-red,   on black stage-floor.
                    Around you, others swirl about
dressed as green algae,
                    like frenetic atoms
            under a microscope in a dimly lit laboratory.
Art mirroring life — reflecting the pandemic on stage.

Your hands palpitate,
                    as the sun's own blinding yellow corona
cracks through the cyclorama.
                People leap about — masked, veiled.
                                You snare a man's sight
with your fingers mimicking a *chakravavyuh* —
                    you are red, he is green, she is blue —
trishanku — life, birth, death —
                    regermination, rejuvenation, nirvana.

Everything on stage — as in life —
                                moves in circular arcs.
Irises close and open, faces veiled unveil —
            hearts love, lungs breathe — breathless.

Lights, electromagnetic —   *knotted, unwrapped* —
                    music pulsates, reaching a crescendo,
                                    then silence.
Time stops. Far away in the infinite blue of the cosmos —
        I look up and spot a moving white.
I see a white feather
                        trying its best to breathe
in these times of breathlessness, floating downwards —

and as it touches the floor, in a split-second
everything bursts into colour, movement, the *bols/taals*
                              try to restore order,
rhythm,   both contained and free.

The backdrop bright orange,
                    the silhouettes pitch-black.
As you embrace another humanform,
          the infinite journey of timelessness might seem
                                        inter_rupted,
but now is the moment to reflect and recalibrate
immersed in the uncharted seas, in the *widening circles*,
                    telling us —   others matter,
the collective counts.

I examine minutely the striated strands
                    of the pirouetting feather, now fallen —
its heart still beating, its blood still pumping,
                              its white untarnished.
Life's dance continues — with or without us —
only in the understanding of what is,
                              is there freedom from what is.

꒒

New Delhi

7
Consolation
HOPE

Love is not consolation. It is light.
—SIMONE WEIL

*... and then, I have nature and art and poetry,
and if that is not enough, what is enough?*
— VINCENT VAN GOGH

*To see a World in a Grain of Sand
And a Heaven in a Wild Flower
Hold Infinity in the palm of your hand
And Eternity in an hour*
— WILLIAM BLAKE, 'Auguries of Innocence'

THE GIFT OF LIGHT

However vast the darkness,
  we must supply our own light.
— STANLEY KUBRICK

Through filigreed *jali* screens
  of pink-ochre sandstone,

light melts to darkness
  in the *sanctum sanctorum* —

Emperor Humayan lies
  entombed in marble.

In this dusk-penumbra,
  another laser-sharp beam

crosscuts the axis
  of the light's path,

its precise angular shaft —
  a dream, an illusion —

its gradation subtly aligned,
  its frequency pitch-perfect.

Our own quiet breathing —
  a silent sacred song

softly muted— unlike the
  azaan's piercing prayer call.

The gift of light
  is life's benediction

in these dark times —
  no matter what or where,

there is always light.

Humayun's Tomb, New Delhi

SUDEEP SEN

BURNING GHATS, VARANASI

for Puneeta & Sanjoy Roy

Over-heated flaming pyres of the burning dead
        partially shield my sight of river Ganges —
its fast muddy currents eddying the floating lamps,
        bathing bodies,
                remains of corpses, flesh-bone ash.

At Manikarnika Ghat, a mixture of sanctity and stench
        rises from silted sands and wooden armatures —
fire-aided decomposition of human flesh —
        the offerings swiftly lapped up by roaming animals.
An emaciated sadhu with wild-knotted dreadlocks,
              perched precariously on a bamboo frame
                    on the edge of the river,
dreams of alms that might come his way,
            even at this late hour.

Presiding priests, feed ritual ghee
        to the burning wood-and-dead —
              the flames forming huge flares,
          fragmented waves of golden-amber spark,
electrifying helical fire-flurries —
            a living, crematorium drama.

A young boy scratches his newly-shaven head,
        a pot-bellied man immerses himself in the river,
stray dogs bark, cows groan, loudspeakers bray.
On ghat-side walls, Gandhi posters preach peace.
        Amid so much noise,
            the business of death being transacted
carries on, without any emotion or fuss.

Saffron-robed men on ghat-steps
                sit in yoga postures, praying —
                  a silent quest —
what does prayer amid all this din and commerce
            get you anyway?

Medley of bells, conch, chant, fire, water, boat, people
        ceases to be a cacophony after a few hours —
variant decibels melding into a drone, a trance —
            where the only balance that exists,
is in our minds.

Bare-headed, bare-bodied young men,
                draped in swathes of pure cotton,
        foreheads smeared in sandalwood and vermillion
            carrying ash-filled earthen pots —
walk past me towards the river-edge,
detached —
        eldest sons performing last rites for their dead.

White-clad teachers   squatting cross-legged on the ghat
            under large circular cane-parasols
    impart teachings from the Upanishads and Vedas
to young priests-in-the-making.

Illuminated cane-lanterns
        hang on long bamboo poles curving skywards —
            homage to the memory of martyrs —
guiding light for heavenly apsaras
descending during the Kartik month to bathe in Kashi —
as oil-soaked wicks flickering on beds of rose petal, sail
                    catching the waves' moods.

In the super-heated pyre, I hear another ritual pot break,
            another skull crack, another soul take flight.
I see some shore-temples slow-sink
                    into the swallowing river —
effects of unpredictable tides and climate change
    taking with them, both the mortal and the immortal —
Holocene's carbon-footprint — its death text, unceasing.
Ashes to ashes, dust to dust —
            water to heavy water, life to after-life.

॰

GANGA, RISING

not an absence but a presence, / dense as any
mineral,.... // ... of consciousness enacting its ...
insurgency / against a dark mountain.
— CAMPBELL MCGRATH, 'Time'

Iridescent turquoise and muddy brown meet,
forming a darker shade of pleated fresh water.
At Devaprayag river junction, Bhagirathi and
Alaknanda merge. At this intersect, Ganga is
born.

On a slippery rock ledge, a sadhu in saffron
robe sits cross-legged in a yoga *asana*,
meditating — two wet oval stones placed atop
each other in front of him are all it takes to
build a sacred place for worship — asymmetry
does not matter.

He prays in silence, the surging rivers chant
in chorus — inner calm, nature's noise. Sparse
paraphernalia of 'stone-water-prayer' —
trimurti's perfect triad — music of the spirit.

For some, it doesn't require much to realise
dreams — a modest yearning, a higher quest
— *trishul* / trident balance held in perpetuity.

Here, there is no space for perfectly rounded
pebbles or gentle musings — only large granite
outcrops can shackle the soul's ferocity — a
jagged fierceness — not harsh, yet quietly
robust.

ASPEN
for Simi

Gold-orange patina
  imprinted serrated leaves

glow silk — incarnadine
  like russet sunsets.

Foliage slow-shivers —
  every breath, heaving.

Winter-white barks
  studiously slow-burn.

Forest fires conflagrate,
  but cannot raze

the incandescent love
  for my beloved.

Wave after wave,
  the Northern lights'

luminously pirouette
  in polar-cooled wind-

chill — redolent colours
  sculpting translucent

letters in this frozen air —
  a sacrament of faith,

brightly lit. Decoding
  hieroglyph's lost

lyrics — an exquisitely
  sung ghazal unfolds.

Glava, Sweden

SHIULI | HARASINGARA
for Simi

October's autumnal month
   splashes white and orange
on the evergreens —

plant that replants
   and transforms itself —
a nuptial hint for raw love.

Delicate soft-white *shiuli*,
   its short saffron-stem —
bleeds passion in the cool air.

Soon the festivities, food,
   flowers, camaraderie,
prayer, will infuse everything —

the weathervane will turn,
   mirroring floret-twirls
and pirouettes in the breeze.

Centrifugal petal patterns
   will reveal micro-matrix —
cosmos' macro design.

Shiuli remains my goddess,
   my lover — *harasingara*
its colour-coded nomenclature.

As I intimately clutch
   fistfuls of sepals, petals —
I remember Banalata, lines

from Andal and Sappho.
   My yearning, untempered;
my memory, flushed.

Patterns on my palms
  inked in floral-blood —
a chrysalis, a poem.

A silent *mantra*
  hummed quietly in private —
a song for my beloved.

New Delhi

LENTICULAR

clouds suck themselves
  invisibly

vacuuming inward —
  inverted cones

like tornado columns
  arrested mid-flight

in their tracks —
  orographic,

lens-shaped — trying
  to maintain form

and structure's algorithm.
  Dreamscapes

melt and mould —
  flying-saucer-like,

evanescent concentric
  layers — peel, unpeel —

swoop up as smoke
  and slow-melting snow.

Secrets stored
  in these barreling

swirls and wisps
  remain — a mystery.

Chameleon
  clouds, change

colour striations
  with the sun's mood —

molecular magic,
  chromatic DNA —

unveiled only
  for the chosen few.

ॐ

INDIAN SKIES: CINQUAIN DIPTYCH

I stare —
    summer's steaming
sky layered thick with fine
dust, quaking in deep heat — where are
    the rains?

The clouds
    appear distant —
a mirage — hunger for
moisture — that's the only yearning —
    downpour.

[NOTE: The American cinquain is an unrhymed, five-line poetic form defined by the number of syllables in each line—the first line has two syllables, the second has four, the third six, the fourth eight, and the fifth two (2-4-6-8-2). They are typically written using iambs.]

metaphor spark, jagged lightning—
    billion light years arrested,
        suspended in space.

an errant form, soil-metal —
    divorced prematurely
        from mother planet.

an electric streak—
    ultraviolet | infrared,
        creates an illusion —

an asteroid, a shooting star —
    disintegrating fiery flourish —
        high-voltage flash.

LISTEN TO THE STARS

… nova // every impulse of light exploding //
from the core / as life flies out of us …
— ADRIENNE RICH, 'Planetarium', *Collected Poems: 1950-2012*

Listen to the stars — far, flung apart — elsewhere,
nowhere, everywhere. An aural orchestra — distant
   pan-flute crackles echoing anti-gravity static,
space-dust murmurations, galactic-sighs, crests-troughs.

What frequency or wavelength powers them?
No sine or cosine graph can map. No telescopic lens
   nor Hubble's metal ears — strong enough to decode
their algorithm, decipher their grammar, text, language.

What we see, we see / and changing is changing //
the light that shrivels a mountain / and leaves
   a man alive // Heartbeat of the pulsar / heart sweating
through my body // The radio impulse / pouring

   in Taurus // I am bombarded   yet I stand ….
A philharmonic score — a nebula poem, a long poem,
   an epic — its unceasing ocular cadence, haunting —
black hole's cosmic crescendo — an axiomatic trance.

🐦

Corona: Elliptical Light

The sun shone, having no alternative, on the nothing new.
— SAMUEL BECKETT, *Murphy*

Early morning light's sluggish haze
  clears in a flash — a cloud break

illumination, so sharp and bright
  it blinds the birds on my neem tree

as they recoil within their feathers.
  Falling on new buds, the ray's glare

splits open their perfect coronas —
  pollen shower-burst, an ochre-flare,

yellow micro-comet in slow motion.
  A solar system, embedded —

sepals, petals, ovule, ovary, stigma, style —
  concentric filament folds, anther rings,

mimicking elliptical planetary orbits
  holding stalk's epicentre intact —

centripetal and centrifugal, balanced —
  an analeptic birthing, an iatric meteor.

🦌

CONSOLATION

Wet rose petals, velvety,
gleam, tealight's soft glow.

Mise-en-scene — dreamy,
partially out-of-focus.

Feather-fire radiates
early seasonal warmth

adding grain-texture
to counter-top's varnish —

unstained walnut wood,
steadfast, frame askance.

❧

Outside in the garden —
night candles, fairylights

on trained greens —
cast striated shadows,

complementing raw
solemnity, soon to follow.

Crystal glass in hand —
hot water, honey, lemon —

a young sufi singer awaits,
his songs slow to unfurl.

❧

In calibrated sonic layers,
he sings raag's incipient

aalaap — a gentle rising
form, its timbre deepening,

measured lyrics unfolding
note by note, its phrasing

spartan, simple, secular,
spiritual — a deep sonar

healing — its soul sombre,
magical, meditative.

Breastfeeding

In the dry heat of the extended pandemic, your lips lacerate my nipples as we nourish each other — mother and child. The weather and the times have not been kind to my health — my skin perpetually dry and wanting. You are helpless in your infant need for food, as I am in my need to satisfy your hunger. My chapped aureolas are in pain, and yet there is love — strange love it is that hurts physically. Love is not a perfect equation, it never was and is unlikely to be — yet we are beholden to it.

for Jatin Das

1.

Pleated, metrical folds
   of dried palm-leaves fan
out in serene oval shapes

gently stirring the still air —
   hands in slow-motion
calibrating power

of a nascent breeze,
   softly soothing
our tropical skins —

trying to assuage the un-
   forgiving unpredictabilty
of climate change.

2.

For now, I am happily
   distracted by my modest
handmade pankhã,

its intricate thread-work,
   its simple woven stories
in natural organic dye —

taking me back
   to childhood memories,
of innocence,

of fair weather —
   and the spare simplicity
of pure clean air.

ॐ

ASH SMOKE

haiku

something still remains —
otherwise from ashes, smoke
would not rise again

8

Lockdown

READING | WRITING

I am losing the habit of speech.
— ROBERT ALAN JAMIESON

The formalization of the vernacular.
— MICHAEL ONDAATJE

HANDWRITING

for Michal Ondaatje

three haiku

fingers hold the pen
firmly, guiding the gold nib
  in *wild cursive scripts* —

lines delicately
etched, perfectly pitched with the
  stylised slant of a

fine and practised hand —
letterforms and words
  bloom, come alive — spell.

Paper T[r]ails

Paper dreams within the cover of a book,
book binds itself with the glue of a spine,

spine weaves together — dovetailed
by the grace of words — words of passion,

words of grief; words of love, hate, wisdom.
Paper crafts its papyrus origins

journeying from tree to table
through clefts, wefts, contours, textures —

transforming from wood to sheet —
white sheets born of unbleached

natural shade — a *tabula rasa* waiting
for ink, graphite, or sable-hair touch.

꒳

Old-fashioned switches — dormant —
now spark static electricity. Paper imagines —

crisp, letter-strewn, bookish, word-wedged.
Phrases elegantly poised, ready to trip off

a palette, exposing photographic plates —
bromide undulations of an untold story —

a narrative to be matted and mounted —
a frame freeing open its borders to dream.

꒳

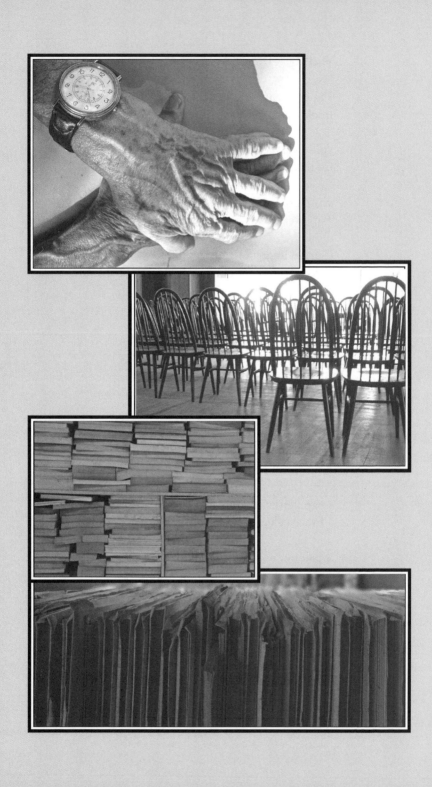

Ilhan's weathered hands, their bulbous veins
hold time and text beautifully phrased —

he is a poet and painter, lover of the sea,
light, silverfish, a sculptor of history.

Like a musician recording his lyrics —
magnetic forces marrying science

and arts — he swims on crest-troughs
of sine-graph modulations, through

physics' precision of arithmetic and tact.
Paper dreams in stacks, between covers,

among notes left surreptiously
between pages for someone else to read.

A stray reader may find the letters —
electric text — unframed and borderless.

Note on the Context:
   'Paper T[r]ails' is a series of tightly wrought images
— set as pastiche montage, sometimes in diptychs and
triptychs — charting 'paper tales' through their journey
of birth, growth and creativity. With subtle use of
natural light and controlled framing, material textures
and contours, lines and phrases from the original poem
as photographic titles, the black-and-white panorama
unravels a narrative that is often hidden to an everyday
eye.
   Empty shelves dream of words they may have
contained, the history of their making, the music that
lies therein, lover's clues to be chanced upon and
uncoded.
   The writer is a poet, the photographer a painter, and
the viewer a lover — he is also the creator, preserver
and destroyer — alluding to the triadic Hindu myth of
Brahma, Vishnu and Shiva.
   However, the palette is secular in nature, precise
like architecture and arithmetic, fluid and spontaneous
like a song and a story.

LANGUAGE

for Anamika, Malashri, Namita & Neeta

Without translation, I would be limited to
the borders of my own country. The translator is
my most important ally.
— ITALO CALVINO

My typewriter is multilingual,
its keys mysteriously calibrating

my bipolar, forked tongue.
Black-red silk ribbon spools, unwind

as the carriage moves right to left.
In cursive hand, I write from left to right.

My tongue was born promiscuous —
speaking in many languages.

My heart spoke another, my head
yet another — the translation, seamless.

❧

Auricles, ventricles pump blood —
corpuscle-like alphabets, phrases, syntax

cross-fertilize my text, breathing life.
Texture enriched — music, cadence

spatially enhanced — osmotic,
polyglottal — a polygamy of grammar.

Letterforms dance, ligatures pirouette —
ascenders, descenders — pitch perfect.

Imagination isn't caged in speech —
speech cannot be caged in language.

❧

FOUNTAIN PEN

for Dinesh Khanna & Rachana Yadav

haiku

tactile pleasure of
a nib slowly caressing
the skin of a page

WRITING

haiku

phrases, words commune —
elliptical — moulding raw
imagination

POETICS OF SOLITUDE, SONGS OF SILENCE
for Pico Iyer

"I find it wholesome to be alone the greater part of the time. To be in company, even with the best, is soon wearisome and dissipating. I love to be alone. I never found the companion that was so companionable as solitude," wrote Henry David Thoreau in *Walden*. Solitude is something most creative writers and artists crave for, and yet when it is forced on to you — how does one cope?

As a poet and literary writer, and as a person who works from home (unless I am travelling on work), 'self-isolation' is nothing new, abnormal or unusual. Over the last three decades, I have spent most of my working hours happily and voluntarily self-isolated and quarantined, cocooned in the world of ideas, surrounded by books and literary artifacts in my office-study. The only ostensible ambient sounds — rustling florets of neem leaves outside, assured metronomic ticking of an antique clock, soothing sounds of running water from a clay Zen water-fountain, familiar scratch of graphite point at the end of my sharpened pencil, and the seamless score created by the soft tap-touch of my fingers on the laptop keyboard.

I have never needed external causes to internalize and live life solitary and indoor. Wherever I am, I'm always at once at 'home' and in the 'world.' Perhaps this ease of simultaneity comes from a sense of rootedness. Like a large banyan tree with tertiary trunks and branches resembling fused stalactites and stalagmites — the veins and arteries of ideas flowing omni-directionally at all times. And yet in this isolation and solitude, there is an inherent yogic sense of centredness, where being with oneself is both wholesome and multitudinous. It is a precious zone for philosophical and creative thinking, a space for silence and "stillness" (as Pico Iyer says) that allows an inner voice to be heard.

The idea of 'white' space has always been important to me, both in my art and in my living. In an increasingly noise-polluted world, it is a space of calm, silence and solitude.

In white we have the entire spectrum of colour and beyond, beyond infrared and ultraviolet. In poetry, art or photography presented on a flat surface — what is left out of the frame is equally important to what is inked-in as words and images. Without the silence of the white space — the work's overall entirety would never be balanced. The visible and the invisible act as a *yin-yang* with a calibrated fulcrum providing the mood and texture to tonality's subtle equilibrium.

This theme often finds expression in my writing, as in my poem, 'Silence', from my book *Fractals*:

> Silence has its own
>   subtle colour.
> Between each breath
>
> pause, heat simmers
>   latent saliva —
> tongue-entwined lisp.
>
> Here and there,
>   errant clouds wait,
> yearning for rain.

Desire melting
   even silence to words —
word's colour bleed

incarnadine, as your lips
   whisper softly
the secrets of your silence.

Your fine chikan blouse —
   white, sheer,
and almost transparent —

cannot hide the quiet
   of your heart-beat
on your wheat-olive skin.

The milk-white flower
   adorning your hair,
sheds a solitary petal,

just one. In that petal
   silence blooms colour —
white, transparent white —

   pure white silence.

I have grown up, worked and lived for many years at a stretch, in some of the busiest cities of the world — Delhi, New York, London and Dhaka. In India's capital city of 26 million and a country that hosts 1.6 billion people — I have got so used to the cacophony and external sounds that I can instinctively tune them out, at will. Whether I am in a packed train, or in a café, or walking on a crowded street, I can — if I choose to — just detach completely and go into my own zone of blur, which within seconds turns to a calming silence. Through years of untutored regimen, this process has become second nature, like any meditative practice.

Albert Camus in *The Myth of Sisyphus and Other Essays* wrote, "In order to understand the world, one has to turn away from it on occasion." At one level, as a poet, one can often feel out-of-sync with the pedestrian conduct of the world at large. Being alone and confined is often a refuge from banality. Whichever way one looks at it, I must confess to being more than a little amused (even though I understand it), to see the world enter the phase of 'social distancing' 'self-isolation' — an idea that some of us have known as a lived reality for a very long time. So one carries on with the day as usual, the week, the month, the rest of the year, and more — without any fuss, distress or alarm.

Throughout history, writers have sung paeans in favour of isolation and solitude. Aldous Huxley: "The more powerful and original a mind, the more it will incline towards the religion of solitude". Albert Einstein: "I live in that solitude which is painful in youth, but delicious in the years of maturity." John Milton in *Paradise Lost:* "Solitude sometimes is best society." Thomas Mann: "Solitude gives birth to the original in us, to beauty unfamiliar and perilous — to poetry."

For me, poetry is omniscient, poetry is life, poetry in its widest sense is a way of living. The unipolar focus on what one is engaged in, both centripetal and centrifugal, is the key — a well-made, well-worn universal key that is robust and resilient enough to open up any vista you can imagine. Ultimately everything begins and ends with the poetics of solitude.

LOSING THE HABIT OF SPEECH, REGAINING THE HABIT OF READING

"I am losing the habit of speech. / Language itself grows strange. / I forget the names of all I knew. / Colour, shape and texture blend. / Those things there, what are they called? / Those people, what is it that they do? / How do those little words go / Sense to make situations of? / Soon, I will only meow / or bark / or quack. / My tweets will all be bird-like." Robert Alan Jamieson in this new poem eloquently sums up what it is like to be in isolation in these times of crisis, how the absence of people forces us to be silent, how the lack of industrial and urban activity allows nature to reclaim its unpolluted spaces.

Apart from my usual reading/writing regimen, I'm also sorting my books/papers — reshelving my extensive library country-wise, genre-wise, alphabetically. One of the joys is rediscovering books I'd liked, with pencil notes that I'd once scribbled, and newspaper review-clippings that I'd inserted between the pages — and if I am lucky enough, even handwritten correspondence with the author.

Amitav Ghosh's *The Great Derangement* (Penguin) tops my list for its urgency and intelligence, for its war cry to tackle what is perhaps the most serious issue that faces our world. Everything that we are going through now is in some way related to anthropocene. Ghosh writes, "Climate change poses a powerful challenge to what is perhaps the single most important political conception of the modern era: the idea of freedom, which is central not only to contemporary politics but also to the humanities, the arts and literature." This must re-read is a wise, elegant book.

Pico Iyer is another favourite writer whose every book I devour. *The Art of Stillness* (Ted) can't be more appropriate. But I opted for his newest, *A Beginner's Guide to Japan* (Penguin), a beautifully articulated fable — its spare haiku-like poetic prose raising the art of minimalism to a new height. As they say in the temples of Kyoto, "The opposite of a great truth is also true." This book is set in six sections with sub-titles such as 'Several Grains of Salt', 'Empire of Smiles', 'The Bridge of Hesitation', and 'The Other Side of Sorrow'. Iyer's quiet and sharp meditations are a perfect tonic for our mind/soul. So unsurprisingly, the book opens with a Simone Weil epigraph: "Attention, taken to its highest degree, is the same thing as prayer."

Siddhartha Mukherjee, at a young age, has contributed significantly to the art of stylish medical/science writing, highlighting the microcosm that bind complex neural energies in our body and mind. People know his gargantuan prize-winning books, *The Emperor of All Maladies,* and *The Gene* — but I urge you to reread *The Laws of Medicine* (Ted), a delicate treatise that packs an equally potent power. Using experiential knowledge and past cases' data, he outlines "three principles that govern modern medicine" — Law One: "A strong intuition is much more powerful than a weak test." Law Two: " 'Normals' teach us rules; 'outliers' teach us laws." Law Three: "For every perfect medical experiment, there is a perfect human bias." This boutique book, quoting

J K Rowling and John Locke to anchor his diary-like journaling and thesis, is beautifully written. And like his big books — this is equally thought provoking and informative, as it is human.

Appropriately, do read Abraham Verghese's 1994 book, *My Own Country* (Phoenix), which is not just an ordinary account of his patients, but a memoir of how his patients changed the way he looked at the disease and life. It is deeply moving and elegantly written.

When I started out writing this piece, I assumed I'd talk about poetry/poets I've admired or rereading — Milton, the Symbolist Poets, Paz, Neruda, Celan, Walcott, Brodsky and Heaney. So I'm surprised that I ended up recommending non-fiction. Perhaps in the above books, the strict dividing lines between fiction, non-fiction and poetry are blurred. And as in any good piece of writing or book — it is the content, the language and the emotional intelligence that ultimately matters.

9
Epilogue
PRAYER

*In prayer it is better to have a heart without words
than words without a heart.*
— MAHATMA GANDHI

Physical matter is music solidified.
— PYTHAGORAS

MEDITATION
seven linked-haiku

Inhale, exhale — eyes
shut. Sound of collective breath —
invisible song.

Thinking of people
one loves and hates — we forgive.
The golden light streams

from the crown of your
head, down your neck, shoulder, spine,
your arms, legs — outward

to embrace one and
all. Under your feet,
the imagined bowl of blue

light sustains you, me,
and us. Long, short, rapid, slow
breathing centres us.

At the end, we rub
our fingers and face, we
open our eyes — and

smile. Eyelids veiling
our pupils, see — a sense
of calm, peace and love.

PRAYER

Prayer flags
    flutter —

I try to catch
    their flight —

their song, their words,
their flap.

They are quiet
    in their colours —

the colours of river,
snow,
    earth, leaves, sun.

They are waves
    that turn the earth,

the earth moves them.

Five colours mix —
      white.

CHANT

om ma ni padme hum

blank contains
        everything —

everything contains
        nothing —

 nothing contains
        all.

all is one —

        one is many —

 many is all.

om ma ni padme hum

ॐ

OM: A CEREMENT

for Carolyn Forché & Joy Harjo

Architecture of frozen music.
— GOETHE

In my city, I am surrounded by constant cries
   of the dying, burning pyres heaving

under burden of wood, smoke and bones —
   wailing summed up by sonic notes of *Om*.

Civilisation's first sound — Sanskrit syllable
   echoing a conch shell's harmonic mapping —

its involute spiral geometry holding within
   and emanating airborne sonar screams.

My ancestors, grandmothers, mother — blew
   into this smooth shell cupped in their palms,

held intimately as if it were a talisman,
   a prayer, a *pranayam* in yoga's daily ritual.

But breathing is a privilege these days —
   pandemic-struck, oxygen-deprived,

my friends perish, the country buckles, airless.
   Even an exquisite cerement lacks the sheen

or wax to wrap the contours of a corpse now.
   Each day as I write endless condolence notes,

etching dirge-like couplets on gravestones —
   my city continues to be dug up — not to make

space for burial sites, but for palaces of illusion:
   an *architecture of frozen music*, greed, calumny.

A country without a government,
    a country without a post-office — Shahid laments:

"Let me cry out in that void, say it as I can.
    I write on that void." *Om's* celebration now

an unceasing requiem. Yet we chant in hope,
    for peace: *Om Shantih, Shantih, Shantih.*

ॐ

My deepest gratitude to the publishers and editors at Pippa Rann Books & Media and Penguin Random House for bringing this book out, and disseminating it with care and enthusiasm.

'Love in the Time of Corona' was first commissioned by *The Indian Express* newspaper and appeared in its weekend magazine section. Subsequently, it also appeared in *ArtVirus* magazine (USA); on *Write Where We Are Now* global anthology (curated by the former UK poet laureate, Carol Ann Duffy) hosted on the Manchester Metropolitan University's website; in *Poetry of the Plague* (Poetry Kit, UK) ed. Jim Bennett, in *Kitaab* (Singapore); *The Bengaluru Review* and *The Dhaka Tribune* newspaper (Bangladesh). This poem in various translations have appeared — in French by Bernard Turle in the *Idees Magazine*; in Spanish by Yuri Zambrano premiered at the 'World Festival of Poetry' on their portal and Facebook Live; in Serbian by Milan Jovic & Gojko Bozovic in *Archipelag* magazine; in Turkish by Erkut Tokman in *Caz Kedisi*; in Persian by Fatemeh Ahmadi & Rosa Jamali in *Jenzar*; in Urdu by Asif Aslam Farrukhi in Pakistan's *Humsub*, in Marathi by Sanket Mhatre in *Lokmat* newspaper; in Bengali by Niladri Mahajan; and more. It also appears in the global anthology *Singing in Bad Times: A Global Anthology of Poetry Under Lockdown* (Penguin Random House); and in the 'Stir' (Mumbai) and 'ArtMatters' (Hyderabad) pandemic fundraising initiatives. The poem was also part of an international renga project 'Airborne Particles': *The Calvert Journal*, UK. ed. Ioana Morpurgo.

Thank you to all the editors and publishers who first commissioned and published the following pieces in various newspapers, magazines, journals, anthologies and books (print & online); and to the producers of television, radio and internet programmes for broadcasting them (some of them in earlier versions):

'i.e. [That Is]' and 'Lenticular': *Open* (2021 New Year Double Issue).

'Speaking in Silence': *The New Humanist* (UK).

'The Third Pole': *The Third Pole* (UK).

'Saline Drip', 'Love in the Time of Corona', 'Obituary', 'Hope: Light Leaks', 'Asthma', 'Quarantine', 'Newsreel': *Beltway Poetry Quarterly* (USA).

'Corona: Elliptical Light': *The Hollins Critic* (USA), and broadcast on the Oxford Bookstores Facebook Live show.

'Hope: Light Leaks' and 'Asphyxia': *Row of Masks* (Romania)

'Black Box': *The Dhaka Review* (Bangladesh)

'Poetics of Solitude, Songs of Silence': *Outlook*.

'Losing the Habit of Speech, Regaining the Habit of Reading': *The Telegraph*.

'Ghalib in the Time of Crisis': *The Hindustan Times*.

'The Legacy of Bones': *Strands and The Hindustan Times*.

'The Room You Grew Up In': *The Hindu*.

'Obituary': on the Oxford Bookstores Facebook Live show, and *The Bengaluru Review* & TMYS Academy portal.

'Hope: Light Leaks': *Strands* and TMYS Academy portal.

'Corona Haiku[5.7.5]' — 'Ventilator', 'Covid 19', 'Sanitiser', 'Lexicon': *The Bengaluru Review*.

'Witherstone', 'Air: Pankha Pattachitra', 'Aspen': *The Bombay Review*.

'Burning Ghats, Varanasi': *Indian Literature* (Sahitya Akademi) and *Hans* (in Hindi translation by Mangalesh Dabral).

'Corona Red' is part of a book-in-progress, 'Small Tales from a Kitchen Table' (with photographer, Dinesh Khanna).

'Preparing for a Perfect Death', 'Roza', 'Quiet', 'Touch of Your Absence': *The Punch Magazine*

'Preparing for a Perfect Death', 'Om: A Cerement', ''Paper T[r]ials': *The Dark Horse* (UK)

❖

Some of these pieces — have recently appeared or are forthcoming —in the following books/anthologies:

EroText (Vintage: Penguin Random House) by Sudeep Sen.

Singing in Bad Times: A Global Anthology of Poetry Under Lockdown (Penguin Random House).

Hibiscus: Poems that Heal and Empower (Hawakal).

Shimmer Spring (Hawakal).

Dusk to Dawn : Poetic Voices on the Current Times (Heritage).

Touched: An Anthology of Voices Writing on Creativity and Madness (Speaking Tiger).

The Yearbook of Indian Poetry in English (Hawakal).

Open Your Eyes: Climate Change Poetry Anthology (Hawakal).

Voices (Jaipur).

Corona Crisis: Writings from the Debris (HarperCollins).

Kritya International Poetry Anthology (Poeisis).

Covid 19 World Anthology of Poetry (Kistrech, Kenya).

Coming out of Isolation (Kistrech, Kenya).

Airborne Particles (The Calvert Journal, UK).

Poetry and Covid (a project funded by the UK Arts and Humanities Research Council, University of Plymouth, and Nottingham Trent University).

A selection of the 'Skyscapes' photographs were first featured as part of the 'Lockdown Diaries' exhibition curated by Aditya Arya for the Indian Photo Archive Foundation at the Museo Camera, Gurgaon.

The photographs accompanying 'Paper Trails' were first featured at the United Art Fair (New Delhi), and are now part of the collection at the Auslandjournal ZDF Television (Berlin).

SUDEEP SEN's [www.sudeepsen.org] prize-winning books include: *Postmarked India: New & Selected Poems* (HarperCollins), *Rain, Aria* (A. K. Ramanujan Translation Award), *Fractals: New & Selected Poems | Translations 1980-2015* (London Magazine Editions), *EroText* (Vintage: Penguin Random House), *Kaifi Azmi: Poems | Nazms* (Bloomsbury), and *Anthropocene* (Pippa Rann). He has edited influential anthologies, including: *The HarperCollins Book of English Poetry, World English Poetry, Modern English Poetry by Younger Indians* (Sahitya Akademi), and *Contemporary English Poetry by Indians. Blue Nude: Ekphrasis & New Poems* (Jorge Zalamea International Poetry Prize) and *The Whispering Anklets* are forthcoming. Sen's works have been translated into over 25 languages. His words have appeared in the *Times Literary Supplement, Newsweek, Guardian, Observer, Independent, Telegraph, Financial Times, Herald, Poetry Review, Literary Review, Harvard Review, Hindu, Hindustan Times, Times of India, Indian Express, Outlook, India Today*, and broadcast on BBC, PBS, CNN IBN, NDTV, AIR & Doordarshan. Sen's newer work appears in *New Writing 15* (Granta), *Language for a New Century* (Norton), *Leela: An Erotic Play of Verse and Art* (Collins), *Indian Love Poems* (Knopf/Random House/Everyman), *Out of Bounds* (Bloodaxe), *Initiate: Oxford New Writing* (Blackwell), and *Name me a Word* (Yale). He is the editorial director of AARK ARTS, the editor of *Atlas*, and currently the inaugural artist-in-residence at the Museo Camera. Sen is the first Asian honoured to deliver the Derek Walcott Lecture and read at the Nobel Laureate Festival. The Government of India awarded him the senior fellowship for "outstanding persons in the field of culture/literature."

My generation was lost. Cities too. And nations.
But all this a little later.
Meanwhile, in the window, a swallow.
— CZESLAW MILOSZ